Zero Trust Architecture Guidance

Rethinking Enterprise Security from the Inside Out

JC Louis-Charles, CISM

The moral rights of the author have been asserted

Copyrights 2026

All rights reserved

Cybersoft Publishing LLC

Fort Washington, MD 20744

Drclaude.net

First Edition January 2026

Foreword

Somewhere in your organization, a credential issued 18 months ago is granting access to a system its owner no longer needs to use. A contractor who left the project in January still has a working VPN token. An application running in a cloud environment your team provisioned during a pandemic sprint is communicating laterally with a database that holds records it was never designed to query. None of these situations triggered an alert. All of them represent exactly the kind of implicit trust that this book was written to dismantle.

I wrote this book because the conversations I keep having with senior leaders follow a pattern that concerns me. The security briefing goes well. The compliance checklist comes back clean. The penetration test produces a manageable findings report. And yet, when a breach finally happens, it moves through the environment with a speed and reach that none of those artifacts predicted. The reason is almost always the same: the architecture was built on assumptions about trust that have long since become invalid, and nobody revisited them because the perimeter still looked intact from the outside.

Zero trust is not a product you purchase or a project you complete. It is an architectural commitment to verifying every interaction, limiting every privilege, and assuming that any segment of your environment may already be compromised. That sounds exhausting. In practice, it is liberating — because it replaces the impossible task of defending an ever-expanding perimeter with the achievable discipline of protecting individual assets, transactions, and identities at the point where they operate.

This book is written for the leader who needs to understand what zero trust demands of the organization — not just the technology stack, but the operating model, the governance structure, the workforce design, and the cultural shift required to sustain it. Every chapter is built around decisions you will actually face: where to start, what to measure, how to sequence investments, and how to keep the program evolving after the initial deployment loses its executive spotlight.

The perimeter is not coming back. The question facing your enterprise is not whether to adopt a zero trust posture, but how quickly you can dismantle the assumptions that are quietly keeping the old architecture in place. This book is designed to help you answer that question with clarity, confidence, and a plan your teams can actually execute.

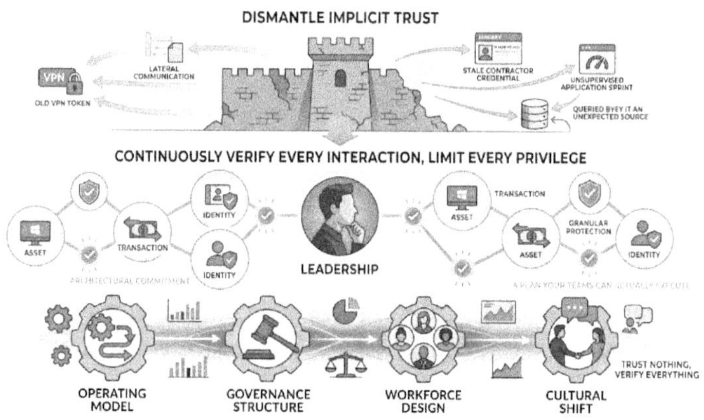

Contents

Introduction

This book is organized around a single premise: zero trust is an organizational transformation, not a technology installation. The nine chapters that follow are sequenced to mirror the decision path a senior leader will follow from the moment the old perimeter model is recognized as inadequate through the point where a zero-trust program is operating, maturing, and sustaining itself across leadership transitions and budget cycles.

The first three chapters establish the foundation. You will begin with a clear-eyed examination of why perimeter-based security collapsed under the weight of cloud adoption, remote workforces, and increasingly sophisticated adversaries. From there, you will move into the governing principles of zero trust — not as abstract philosophy, but as a set of enforceable design constraints that will shape every architectural decision in the chapters ahead. The third chapter maps the technical components — control planes, policy engines, enforcement points — so you can engage your engineering teams with a shared vocabulary and realistic expectations.

Chapters four through six address the core domains in which zero trust architecture must be implemented: identity governance, network microsegmentation, and data protection. Each chapter treats its domain as a standalone design challenge while connecting it to the broader architecture. You do not need to implement them in the order presented, but the sequence reflects the dependency chain that most organizations encounter in practice.

The final three chapters shift from design to execution. The book covers phased rollout with maturity benchmarks, team design and roles, managing cultural resistance and stakeholder alignment, and concludes with a chapter on ongoing governance to sustain the program beyond initial deployment.

Read this book in sequence the first time. After that, use individual chapters as reference points when specific decisions arise. Each chapter is designed to stand alone for the leader who needs to revisit a particular domain without having to reread the entire book. Where one chapter depends on concepts introduced earlier, a brief reference points you back to the relevant section. There are no appendices, reference lists, or supplementary materials to chase. Everything you need to act on is contained within the chapter where it belongs.

1 The Collapse of the Castle Wall: How Perimeter Thinking Broke Enterprise Security

It begins with a routine help-desk ticket. A user in the Chicago office reports that their VPN client keeps disconnecting. The help desk resolves the issue in eleven minutes by resetting a session token. Two days later, the security operations center notices that a file server in the Atlanta data center is staging an unusually large archive of customer records. The investigation reveals that the Chicago session token was harvested by a phishing kit deployed three weeks earlier. The adversary used it to authenticate as a legitimate employee, traversed the internal network without triggering a single firewall alert, and spent seventeen days mapping the environment before accessing anything of value. The perimeter held the entire time. The castle wall never fell. The attacker walked through the front gate.

This scenario repeats itself across industries and organization sizes because it is not a story about a sophisticated zero-day exploit. It is a story about architectural assumptions that had long since become invalid by the time the attacker arrived. The perimeter model assumes that the boundary between inside and outside is meaningful — that traffic originating from within the corporate network is inherently more trustworthy than traffic arriving from outside it. By the time most organizations discovered how badly that assumption had aged, the threat landscape had already outpaced their defensive posture by a decade.

For a manager, the failure of perimeter security is not primarily a technical problem. It is an organizational risk that manifests in financial, regulatory, and operational terms. A breach enabled by lateral movement — the kind the perimeter model cannot stop — typically remains undetected within the environment for weeks or months. Every day of that dwell time represents an expanding exposure: more data accessed, more systems cataloged by the attacker, and more potential for disruptive exfiltration or ransomware deployment. The costs that follow — incident response, regulatory notification, legal liability, reputational repair — are not the cost of a single security failure. They are the accumulated cost of an architecture that was never designed to contain what got inside.

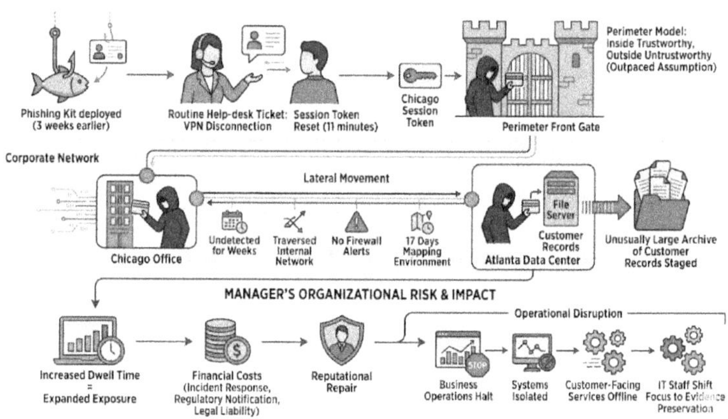

The workflow-level impact is equally significant. When a perimeter breach escalates to an incident response engagement, it does not stay contained to the security team. Business operations halt while systems are isolated.

Customer-facing services go offline while forensic teams determine the blast radius. IT staff who were managing infrastructure suddenly shift to evidence preservation and breach notification timelines. These disruptions have real costs for productivity, customer trust, and market position that no compliance framework can fully capture. Mission-aligned security leadership means understanding this chain of consequences before the incident, not documenting it afterward.

1.1 From Fortresses to Fog: The Architectural Assumptions That Aged Poorly

For the better part of three decades, enterprise network architecture was built around a hub-and-spoke model anchored by the corporate data center. Users sat at desks connected to the corporate LAN. Applications ran on servers inside that LAN. The firewall at the network perimeter examined inbound and outbound traffic, blocking what looked malicious and allowing what looked legitimate. This model worked when the assumptions underlying it held: that users were in fixed locations, that applications lived in known data centers, and that anything inside the perimeter could be trusted because it had already been checked at the gate.

Each of those assumptions began eroding in the mid-2000s and collapsed entirely over the following fifteen years. Cloud adoption moved applications off on-premises infrastructure and into environments that exist entirely outside the perimeter. Mobile work and then pandemic-driven remote work permanently pushed users beyond the firewall. Third-party integrations, API ecosystems, and SaaS platforms created hundreds of trust relationships that the perimeter architecture was never designed to govern. By 2020, the average enterprise had more data and more users operating outside its traditional perimeter than inside it — yet most organizations were still running security models designed for 2005.

1.1.1 Static Perimeters in a Borderless Workforce

A static perimeter is one whose boundaries are defined by physical or logical network topology — the firewall sits at a fixed point, rules are configured for a known set of IP ranges, and the implicit assumption is that the environment those rules protect does not change faster than the rules can be updated. That assumption was never entirely valid, but it was workable as long as the primary computing assets remained in a single data center and the primary workforce worked from company-owned office space.

The shift to cloud infrastructure fundamentally violated this assumption. When workloads migrate to AWS, Azure, or Google Cloud, they acquire IP addresses from provider address pools that change dynamically. When users work from home, coffee shops, or client sites, they connect from IP addresses that the perimeter firewall has no reason to trust. When a business acquires another company and needs to integrate its systems, the acquired environment arrives with its own network topology that may not align with the acquiring firm's firewall rules. The static perimeter becomes a fiction — a set of rules that describe an environment that no longer exists.

The manager's lens here is straightforward: if your organization has more than a handful of remote workers, any active cloud workloads, or any third-party integrations, your perimeter is already porous. The question is not whether to address that porosity but how quickly your architecture can catch up to the reality your workforce has already built.

Diagram 1.1 – Evolution from Static Perimeter to Borderless Architecture

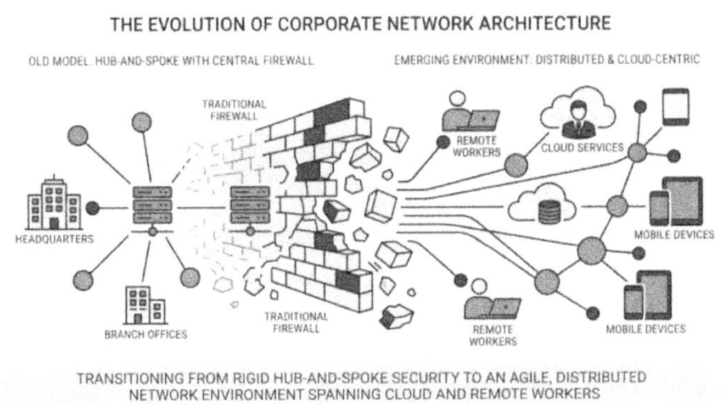

1.1.2 Implicit Trust as a Systemic Vulnerability

Implicit trust is the architectural decision to grant access based on network location rather than verified identity and context. In a traditional perimeter model, a device that authenticates to the VPN is trusted to reach any internal network resource that is not explicitly blocked. A service account created to run a scheduled job has persistent access to the databases it was initially configured to query and often several others until someone manually revokes those permissions. A contractor who completes their engagement retains their credentials until the deprovisioning process catches up with the offboarding process.

These implicit trust relationships accumulate over time until the total access granted across the environment vastly exceeds what any operational review ever intended. Security teams sometimes call this privilege sprawl: the slow expansion of access rights that occurs as systems are provisioned, as staff change roles, and as exceptions are made to accommodate workflow needs and never revisited. Each grant may be justifiable at the moment it is made. In aggregate, they create a map of lateral movement pathways that an attacker can traverse once they have obtained a single valid credential.

Implicit trust is not a configuration error. It is a design choice baked into the foundation of perimeter-based architecture. Addressing it does not mean patching the current system; it means replacing the design principle with a more rigorous one.

1.1.3 How Lateral Movement Turned Internal Networks into Attack Highways

Once an attacker obtains a valid credential and authenticates to the environment — whether through phishing, credential stuffing, or purchasing stolen credentials from a dark-web marketplace — the flat internal network of a perimeter-based architecture becomes an attack highway. No checkpoint within the perimeter requires re-verification of the authenticated session's legitimacy before accessing adjacent systems. There is no policy enforcement point checking whether this identity should be allowed to reach this resource at this time. The firewall, having already confirmed that traffic originated from inside the perimeter, does not ask again.

Lateral movement exploits this absence of internal controls. An attacker who compromises a low-privilege endpoint begins querying the network to discover what other systems are reachable. They elevate privileges by finding and exploiting credential caches, misconfigured service accounts, or unpatched local vulnerabilities. They move from the initial point of compromise toward higher-value targets — the domain controller, the database server, the file share containing sensitive data. This process can take hours or days, depending on network complexity and attacker sophistication, and it leaves traces that flat-network monitoring architectures are poorly equipped to detect in real time.

From a manager's perspective, lateral movement represents the period during which you have the most to lose and the least visibility. The attacker is active in your environment, expanding their footprint, and your detection mechanisms — if they are anchored to perimeter events — are largely silent. Reducing the attack highway requires eliminating the flat network and the implicit trust relationships that make it navigable.

1.2 The Threat Landscape That Outpaced the Defenses

The failure of perimeter security did not occur in a vacuum. It occurred against a backdrop of adversary techniques that evolved specifically to exploit the architectural weaknesses described in the previous section. Criminal organizations, nation-state actors, and opportunistic hackers refined their approaches over years of operational experience, learning what worked and building industrial-scale infrastructure to deploy it at volume. The defenses enterprises invested in were often effective against threats from five years prior, while remaining blind to techniques that had become standard practice.

Diagram 1.2 – Modern Threat Landscape Against Legacy Perimeter Defenses

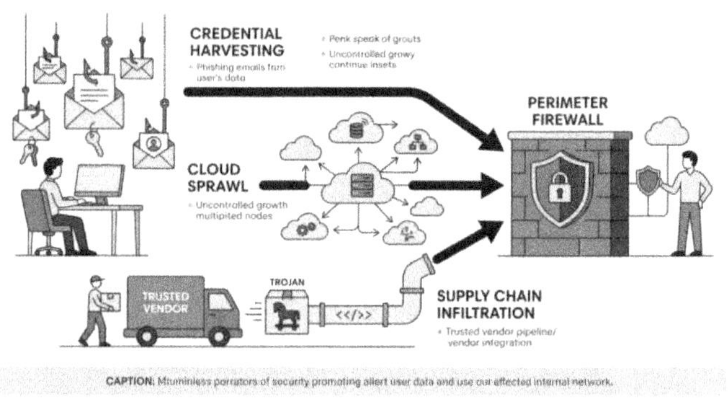

CAPTION: Mtuminless porrutors of security promoting allert user data and use our affected internal network.

1.2.1 Credential Harvesting and the Insider-Equivalent Attacker

Credential harvesting encompasses any technique by which an attacker obtains valid authentication credentials for an enterprise system without directly exploiting a technical vulnerability. Phishing emails that direct users to convincing login replicas capture credentials at the source. Password spraying attacks test common passwords against large lists of known usernames, succeeding against accounts that were never enrolled in multi-factor authentication. Credential stuffing attacks take username-and-password pairs from one data breach and test them against other services, exploiting the pervasive human habit of password reuse. Each of these techniques bypasses the perimeter entirely by presenting the gate with a valid credential rather than attempting to break down the gate.

An attacker operating with harvested credentials presents as an insider-equivalent threat. From the perimeter's perspective, they are indistinguishable from the legitimate user whose credentials they are using. They authenticate through the same VPN gateway, access the same internal systems, and generate the same network traffic patterns — at least initially. Detecting the difference requires looking beyond authentication success to behavioral context: is this user accessing resources at an unusual time? From an unusual location? With an unusual sequence of queries? These behavioral signals are exactly what a perimeter-centric security model is not designed to collect or evaluate.

The manager's response to credential harvesting is not to assume that authentication can be made perfect. It is to architect the system so that successful authentication does not automatically grant the broad access that makes credential theft catastrophic. That architectural discipline — continuously verifying identity, granting access based on context, and limiting the blast radius of any single compromised credential — is the practical foundation of zero trust.

1.2.2 Cloud Sprawl and the Disappearing Network Edge

Cloud sprawl describes the condition in which an organization's cloud footprint grows faster than its governance structures can keep pace. It begins legitimately: a development team provisions a cloud environment to accelerate a product launch. A business unit deploys a SaaS analytics platform to avoid delays caused by internal procurement. A vendor integration requires a cloud-hosted API endpoint that IT is asked to configure over a weekend. Each decision is individually defensible. Collectively, they produce an enterprise technology estate spanning dozens of cloud accounts, hundreds of SaaS applications, and thousands of workload configurations — most of which lie outside the perimeter firewall's visibility.

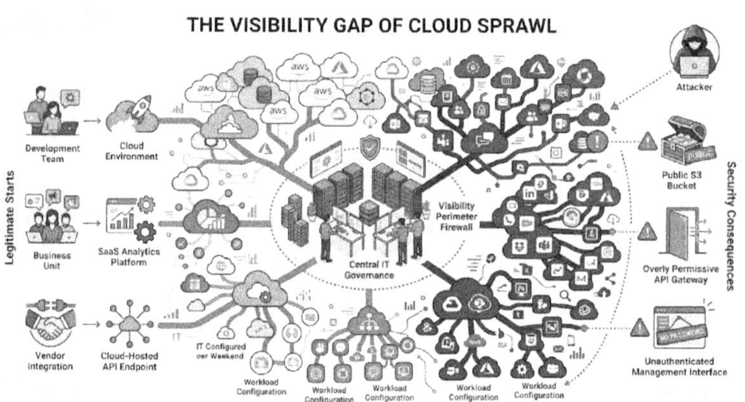

The security consequence of cloud sprawl is not just an expanded attack surface. It is the creation of data stores, compute environments, and authentication systems that operate on their own trust models, issue their own credentials, and communicate through channels that the enterprise perimeter was never configured to monitor. When those environments are misconfigured — a public S3 bucket, an overly permissive API gateway, an unauthenticated management interface — the attacker does not need to penetrate the perimeter at all. The vulnerability is already on the public internet.

Managing cloud sprawl requires governance structures that treat cloud provisioning as a security event from the first moment of deployment, not as a development activity that security reviews later. That governance shift is an organizational decision, not a technical one, and it requires managers to define and enforce provisioning standards across business units that have learned to move faster by operating outside IT oversight.

1.2.3 Supply Chain Infiltration as Perimeter Bypass

Supply chain attacks target the trusted relationships between an enterprise and its software vendors, managed service providers, and integration partners. Rather than attacking the enterprise directly, the adversary compromises a vendor whose code or services the enterprise has already granted elevated access. The enterprise's perimeter defenses, having established trust with the vendor's systems, allow the malicious code or the attacker's activity to pass without challenge.

The pattern is not new, but its scale expanded dramatically when it was used to compromise widely deployed infrastructure software. Organizations that had invested heavily in perimeter defenses found themselves exposed because the software they trusted to manage their systems had been turned against them. The vendor's signing certificate was valid. The update process was legitimate. The malicious payload was delivered through channels explicitly configured by every enterprise security policy to permit it.

From a manager's standpoint, supply chain infiltration demonstrates that trust in a vendor cannot substitute for verifying the vendor's behavior within the enterprise environment. A zero trust posture limits the damage from supply chain compromise by ensuring that even trusted software cannot traverse the environment without encountering policy enforcement points that evaluate the behavior of each session, not just the identity of the initiating entity.

Diagram 1.3 – Supply Chain Infiltration Pathway

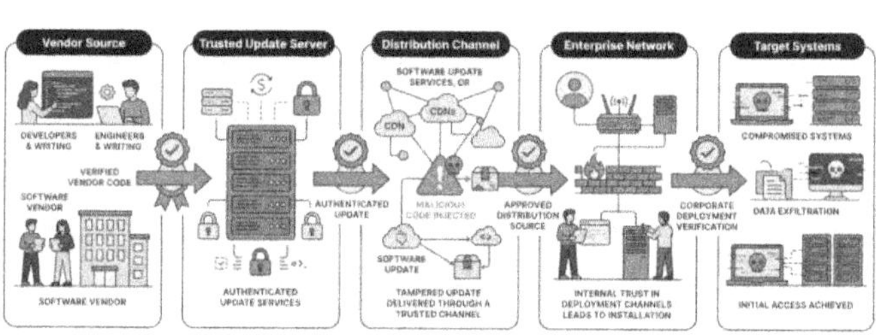

1.3 Quantifying the Cost of Misplaced Trust

Security architecture decisions have financial consequences that unfold over years. The cost of implicit trust in a perimeter-based model is not paid all at once — it accumulates slowly through the period between the initial compromise and its eventual detection, and then all at once when the incident response engagement begins. Understanding these costs in concrete terms is essential for the manager who needs to justify the investment in architectural transformation before the breach rather than after it.

1.3.1 Dwell Time and Detection Gaps in Flat Networks

Dwell time is the period between the moment an attacker first establishes a foothold in an enterprise environment and the moment that presence is detected and contained. Industry measurements have consistently placed median dwell time in the range of weeks to months, with significant variance toward the longer end in organizations with immature detection capabilities. During that window, the attacker is not idle. They are conducting reconnaissance, escalating privileges, establishing persistence mechanisms, and identifying the data or systems that represent their ultimate objective.

Flat networks — those in which authenticated users can reach most or all internal resources without encountering additional access controls — amplify the consequences of extended dwell time. An attacker with weeks of undetected access in a flat network can map the entire environment, identify every high-value target, and position exfiltration or ransomware infrastructure before the security team has any indication of their presence. The detection gap is not simply the period during which the attacker is invisible — it is the period during which the attacker is building toward maximum impact.

Reducing dwell time requires detection capabilities that operate inside the network, not just at its perimeter. It requires behavioral analytics to surface anomalous patterns in authentication, data access, and lateral communication before they reach a threshold visible to traditional alerting systems. And it requires a network architecture that limits how far an attacker can travel during the period between initial compromise and detection — a property that zero trust micro-segmentation is specifically designed to provide.

1.3.2 Regulatory and Reputational Fallout from Perimeter Breaches

A perimeter breach that results in unauthorized access to regulated data triggers obligations that impose direct financial costs on the organization. Breach notification requirements under HIPAA, GDPR, state-level privacy laws, and sector-specific regulations mandate disclosure to affected individuals, regulatory bodies, and, in some cases, law enforcement within defined timeframes. Meeting those timelines requires forensic analysis sufficient to establish the scope and nature of the breach; work that consumes significant internal and external resources in the immediate aftermath of detection.

Regulatory penalties for inadequate security controls or delayed notification can be substantial. But the costs that are harder to quantify and often larger in total are the reputational consequences: customer churn, diminished trust from partners, reduced attractiveness to recruits, and the persistent shadow that a publicly disclosed breach casts over subsequent contract negotiations and procurement evaluations. These costs accrue over quarters and years, not just in the immediate incident window.

For the manager, the regulatory and reputational calculus reinforces a core message: the cost of perimeter-based security failure is not bounded by the cost of the incident response engagement. It extends into future revenue, future partnerships, and future regulatory relationships. Architectural decisions made today including the decision to continue relying on implicit perimeter trust carry cost implications that outlast any single budget cycle.

1.4 The Paradigm Shift: From Border Control to Continuous Scrutiny

Recognizing the failure of perimeter-based security does not automatically produce an alternative. The transition from implicit trust to continuous verification required a conceptual reframe that the industry worked toward for years before it converged on the principles and terminology that define zero trust today. Understanding that reframe not just the conclusion it reached, but the logic that produced it helps managers evaluate architectural proposals and vendor claims against a principled standard rather than a marketing checklist.

Diagram 1.4 – Paradigm Shift: Perimeter Trust vs. Continuous Verification Model

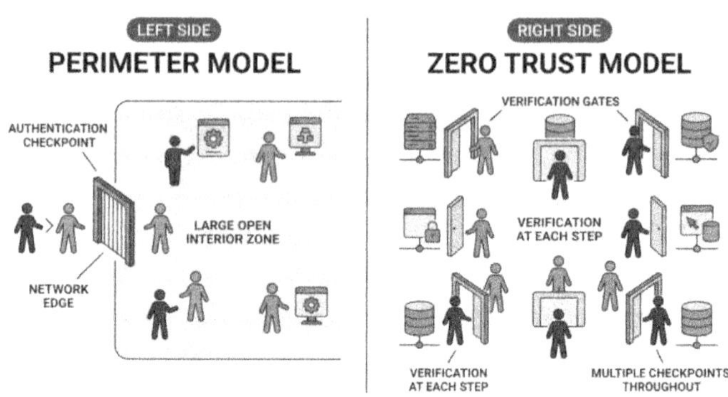

1.4.1 Why the Industry Landed on "Never Trust, Always Verify."

The phrase "never trust, always verify" did not emerge from a single research paper or a standards body. It crystallized from a decade of operational experience with the consequences of implicit trust. Security practitioners who had watched credential theft enable insider-equivalent attacks, observed lateral movement traverse flat networks unchecked, and seen supply chain compromises exploit trusted software channels arrived independently at the same conclusion: the assumption of trustworthiness based on network location was not a defensible default.

"Never trust, always verify" reframes the default assumption. Instead of asking "is this traffic inside the perimeter?" a question that yields a binary, increasingly meaningless answer, it asks: "is this identity, on this device, in this context, requesting this resource, at this time, consistent with established policy?" A perimeter firewall cannot answer that question because it lacks access to most of the relevant signals. It can only be answered by a system that aggregates identity, device posture, behavioral history, and resource sensitivity into a continuous access decision.

For a manager, "never trust, always verify" is not a slogan to put on a slide. It is an organizational commitment to build and operate the systems required to answer that more complex question continuously, at every access event, across every resource, and to make access decisions based on the answer.

1.4.2 Early Adopters and the Lessons They Documented

The earliest large-scale implementations of what would become zero-trust architecture emerged from organizations with firsthand experience of the limitations of perimeter security and the engineering resources to build alternatives. These implementations were not labeled "zero trust" at the time; the terminology had not yet stabilized, but they shared a common architectural direction: eliminate implicit network trust, require continuous verification of identity and device posture, and enforce access policy at the resource level rather than at the network boundary.

The lessons these early adopters documented are valuable not because they describe products that can be purchased off the shelf, but because they articulate the design principles that made their approaches work. The consistent findings were that eliminating the privileged internal network did not significantly degrade user productivity when identity-based access was implemented properly; that device health signals were essential to meaningful access decisions; and that the transition required sustained organizational commitment rather than a single deployment project.

These documented experiences provided a practical foundation for the standards and frameworks that followed, and they remain useful reference points for organizations evaluating what zero trust can realistically achieve at scale in production.

1.4.3   Defining the New Security Contract Between Users, Systems, and Administrators

The shift from perimeter security to zero trust represents a renegotiation of the implicit contract that governs how users interact with enterprise systems. Under the old contract, users were asked to authenticate once, typically at the VPN gateway or the office workstation, and were then granted broad access to internal resources without further challenge. The security burden fell on the perimeter, and users experienced relatively little friction once inside it.

The new contract asks users to participate more actively in their own security. Authentication events become more frequent. Device health requirements become explicit conditions for access. Some resources require step-up authentication when risk signals indicate a session change. Users who work from unusual locations, on unmanaged devices, or outside their established behavioral patterns will encounter additional verification requirements. This is not security theater — it is a direct consequence of taking verification seriously in the absence of a trusted perimeter.

For managers, the new security contract has a communication dimension that is easy to underestimate.

Users who encounter additional friction without understanding why tend to find workarounds that create new vulnerabilities. Employees who understand that device health requirements protect both the organization and their own credentials are more likely to comply with enrollment requirements and less likely to resist policy enforcement. The security contract, like any contract, works better when both parties understand what they are agreeing to and why.

Diagram 1.5 – The New Security Contract: Stakeholder Roles in a Zero Trust Model

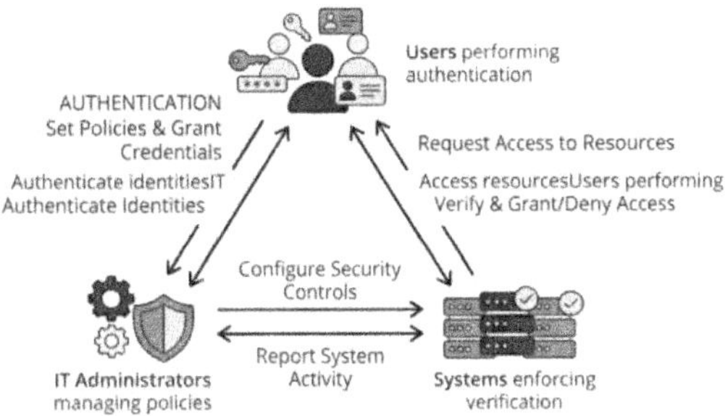

1.5 Manager's Checklist

Audit your current authentication model: determine how many internal resources a user authenticated to the VPN or the corporate network can access without additional verification.

Map your cloud footprint: identify every cloud account, SaaS platform, and API integration your organization has provisioned, including those managed by business units outside central IT.

Review contractor and vendor credential lifecycles: confirm that credentials issued to third parties are time-bounded and deprovisioned through a verified offboarding process.

Assess your dwell-time detection capability: determine how your security operations center would detect an authenticated user behaving anomalously within the internal network.

Identify your three highest-risk implicit trust relationships: start with those where flat-network access creates the greatest exposure to lateral movement toward sensitive data or critical systems.

Communicate the new security contract to your leadership team: ensure executive sponsors understand that zero trust will change user-facing authentication experiences and that the resulting friction is intentional.

Review your regulatory notification obligations: confirm that your incident response plan includes breach notification timelines for every jurisdiction in which you handle regulated data.

1.6 What This Chapter Establishes

Perimeter-based security failed not because the technology was poorly executed, but because the architectural assumptions on which it rested became invalid. Cloud adoption, remote work, and sophisticated adversary techniques each eroded the premise that network location is a meaningful proxy for trustworthiness. The result was an architecture that looked intact from the outside while adversaries traversed it freely on the inside.

The costs of that failure extend well beyond the immediate incident response. Extended dwell time in flat networks allows adversaries to reach maximum impact before detection. Regulatory obligations and reputational consequences persist long after the breach is contained. And the decision to continue patching a fundamentally compromised architecture rather than replacing it accumulates risk with every passing quarter.

The paradigm shift to zero trust, anchored in continuous verification, explicit trust decisions, and internal enforcement, is not a vendor initiative or a compliance requirement. It is an architectural response to a threat landscape that perimeter thinking was never designed to handle. The chapters ahead build the framework for executing that response in a real enterprise environment, starting with the governing principles that give the architecture its internal logic.

Diagram 1.6 – Chapter Summary: From Perimeter Failure to Zero Trust Foundation

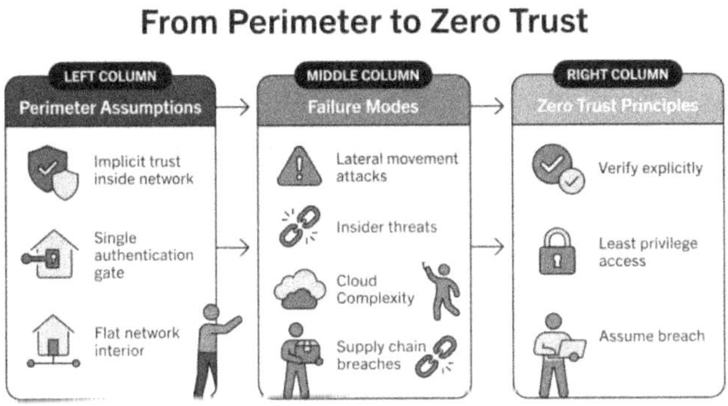

From Perimeter to Zero Trust

2 Never Trust, Always Verify: The Governing Principles of Zero Trust Architecture

A financial services firm completes a six-month zero-trust initiative. The CISO presents a slide showing that 100% of applications are now protected by multi-factor authentication. The compliance team certifies the deployment against the organization's zero trust policy framework. Leadership declares the program complete. Three months later, a penetration test reveals that an authenticated user can still access any internal application from any managed device, regardless of device health status, at any time of day, from any location. The MFA requirement is satisfied at login. No verification occurs afterward. The firm has installed a more sophisticated gate while leaving the interior of the castle just as flat as it was before. This is zero trust theater, the vocabulary adopted, the principles abandoned.

Understanding why this outcome is possible requires going deeper than the surface language of zero trust to the governing principles underneath it. "Never trust, always verify" is not a statement about authentication products. It is a commitment to three interlocking design disciplines: explicit verification at every access event, least-privilege access dynamically scoped to operational context, and an organizational posture that assumes breach has already occurred. Each discipline has specific architectural and operational implications. All three must be honored simultaneously for the approach to provide the protection it promises.

For a manager, the principles of zero trust serve a practical function beyond their conceptual elegance: they provide a stable evaluation framework that is not tied to any specific product or vendor. Technology changes. Threat techniques evolve. Vendor marketing adopts whatever terminology is currently resonant. The principles do not change; they describe what a security architecture must accomplish, regardless of how it accomplishes it. A manager who understands the principles can evaluate whether a proposed deployment actually satisfies them or merely gestures in their direction.

The workflow-level impact of each principle is substantial. Explicit verification changes how authentication infrastructure is designed and how access events are logged. Least-privilege access changes how provisioning workflows operate and how role definitions are maintained. Assume breach changes how detection systems are instrumented and how incident response playbooks are written. None of these changes is purely technical; they require decisions that managers must make, fund, and sustain.

2.1 Explicit Verification: Removing the Free Pass for Network Location

Explicit verification is the principle that every access request to every resource, by every identity, must be evaluated against a defined set of criteria before access is granted. The evaluation is not a one-time gate at the edge of the network. It occurs at the resource level, at the moment of access, using the full set of signals available to characterize the identity, device, location, and behavioral context of the requesting party. Passing a VPN authentication check is not sufficient. Being inside the corporate network is not sufficient. Having passed explicit verification yesterday is not sufficient.

The operational implications of this principle are significant. An architecture built for explicit verification must be capable of collecting the relevant signals, aggregating them into a coherent risk assessment, and applying that assessment to an access decision in real time without introducing latency that degrades user experience or operational efficiency. That capability cannot be bolted onto a perimeter-based architecture. It requires purpose-built identity infrastructure, device management platforms, and policy enforcement mechanisms that operate at the application layer rather than the network boundary.

Diagram 2.1 – Explicit Verification Signal Flow

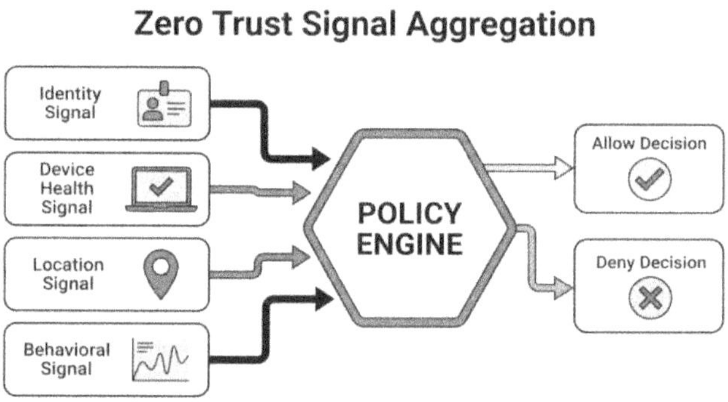

2.1.1 Signal Aggregation: Identity, Device, Location, and Behavior

An explicit verification system is only as good as the signals it evaluates. A system that checks only username and password is performing authentication, not explicit verification in the zero trust sense. A system that checks identity, device health, network location, and behavioral baseline against established policy is performing the kind of continuous, context-aware evaluation that the principle demands.

Identity signals include the authentication credential presented, the identity provider that issued it, the strength of the authentication method used, and the history of authentication events associated with the account. Device signals include whether the device is enrolled in a mobile device management system, whether its operating system and security software are up to date, whether it has passed recent compliance scans, and whether it has been flagged for suspicious behavior. Location signals include the IP address range from which the request originates, the geographic location inferred from that address, and whether the location is consistent with the user's established access pattern. Behavioral signals include the time of day, the sequence of resources accessed, the volume of data requested, and deviation from established baselines.

No single signal is sufficient on its own. A valid credential from an unmanaged device accessing sensitive data from an unusual location at three in the morning is a different risk profile than the same credential from a compliant device in the user's home city during business hours. Explicit verification requires the infrastructure to collect all of these signals and the policy engine to weigh them appropriately in the access decision.

For managers, signal aggregation has a practical procurement implication: a zero trust deployment that lacks device management integration or does not ingest behavioral analytics cannot perform genuine explicit verification, regardless of what marketing materials claim. Evaluating vendor solutions against the completeness of their signal collection, not just the sophistication of their authentication UI, is essential to avoiding the zero-trust-theater outcome.

2.1.2 Continuous Authentication Beyond the Login Event

Traditional authentication models treat the login event as a discrete transaction: the user presents credentials, the system validates them, a session token is issued, and the session continues until it expires or the user logs out. Explicit verification in a zero trust model extends authentication beyond the login event into the session itself. Rather than issuing a token that is valid for a fixed duration, the verification system continuously reassesses the risk profile of the ongoing session based on evolving signals.

Continuous authentication does not require users to re-enter credentials at every interaction — that would be operationally unsustainable. Instead, it operates by monitoring the behavioral and contextual signals associated with an active session and triggering step-up verification when those signals shift in ways that suggest elevated risk.

A session established from a corporate office on a managed device during business hours does not require step-up authentication as long as the signals remain stable. In the same session, if the device location abruptly changes or data access patterns deviate sharply from baseline, a re-verification challenge is triggered before anomalous access is permitted to continue.

From a manager's perspective, continuous authentication is the mechanism that closes the gap between the authentication event and the end of the session the gap that the financial services firm in the opening scenario left unaddressed. It is also the mechanism that makes credential theft significantly less valuable: even if an attacker obtains a valid session token, their behavior within the session will eventually diverge from the legitimate user's baseline, triggering re-verification that they cannot pass.

2.1.3 Risk-Adaptive Access: Tightening Controls When Signals Shift

Risk-adaptive access is the operational expression of explicit verification: the system does not apply a single fixed policy to all access requests from all identities in all contexts. Instead, it applies policies calibrated to the risk level of each request, tightening or relaxing controls as underlying risk signals change. A user accessing a low-sensitivity internal application from a managed device during business hours might satisfy policy requirements with a single authentication factor. The same user accessing a high-sensitivity financial system from an unfamiliar location outside business hours would be subject to additional verification before access is granted.

Risk-adaptive access requires policy definitions that are specific enough to be actionable. A policy that says "sensitive resources require strong authentication" is insufficient if it does not define what constitutes a sensitive resource, what constitutes strong authentication, and what contextual factors might elevate a resource from standard to sensitive classification. Building those definitions requires collaboration between security teams, application owners, and the business units that own the data being protected, a governance process that is organizationally demanding but essential to the principle's operational integrity.

The manager's lens on risk-adaptive access is one of calibration and maintenance. The initial policy definitions are not the end of the work; they are the beginning. As the threat landscape evolves, as user behavior patterns shift, and as new resources are added to the environment, policy definitions must be updated to remain accurate. A risk-adaptive access system running on stale policies is not performing explicit verification it is performing verification against an outdated threat model, which is a different kind of failure.

2.2 Least-Privilege Access as a Living Policy, Not a Static Setting

Least-privilege access is the principle that every identity — whether human or machine — should have exactly the permissions required to accomplish the task at hand, and no more. In a zero trust architecture, least privilege is not a one-time provisioning decision. It is a continuously maintained posture that adjusts as tasks change, as roles evolve, and as the operational context of each session is evaluated. An access right that was appropriate yesterday may not be appropriate today if the task that justified is complete.

The challenge of implementing least privilege in practice is that most organizations have accumulated access rights over the years of normal operations without a systematic process for reviewing or revoking them. Users who change roles retain permissions from previous roles. Service accounts provisioned for specific integrations accumulate additional permissions over time as new dependencies are added. Administrative accounts created for emergency access remain active long after the emergency is resolved. The result is an environment in which the actual access picture what each identity can reach bears little resemblance to the access picture that any intentional policy would produce.

Diagram 2.2 – Least-Privilege Access: From Static Provisioning to Dynamic Scoping

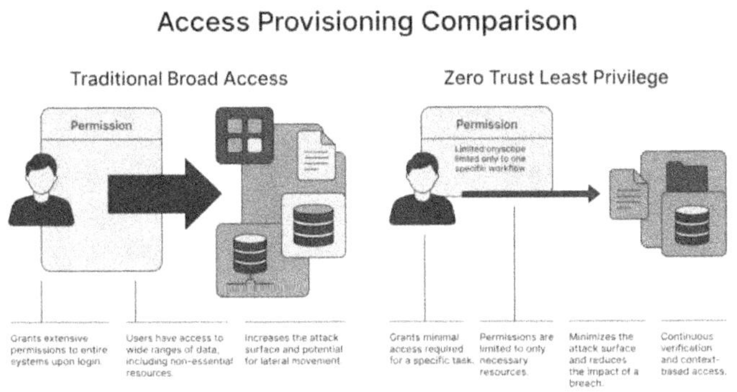

2.2.1 Just-in-Time and Just-Enough-Access Models

Just-in-time access is an operational practice of granting elevated permissions only when needed for a specific task and automatically revoking them when the task is complete or the session ends. A system administrator who needs to apply a patch to a production database does not have standing administrative access to that database they request elevated access, which is evaluated against policy. It may require approval, access is granted for a defined window sufficient to complete the task, and access is revoked when the window closes or the task is confirmed complete. The standing administrative credential that could be stolen and abused does not exist.

Just-enough-access extends the just-in-time concept by also scoping access to what the task requires. A user who needs to read a financial report does not receive write access to the underlying data system. A developer who needs to debug a production issue receives access to logs but not to the production database. Just-enough-access requires task-level policy definitions that most organizations have not historically maintained, and building them requires engagement with the business units and application owners who understand what each task actually requires.

The operational benefit of just-in-time and just-enough-access is not just reduced privilege sprawl. It is a dramatically reduced attack surface for any compromised credential. An attacker who obtains a credential that does not carry standing administrative access cannot escalate privilege through that credential alone — they must also compromise the approval workflow or the policy enforcement system, each of which represents an additional barrier that consumes attacker time and generates additional detection signals.

2.2.2 Scoping Permissions to Workflow Context

Workflow context is the operational state that should govern access policy: what is this identity trying to accomplish right now, and what permissions does that specific task require? In a zero trust, least-privilege model, the access policy is not evaluated in isolation from the work being done. It is evaluated against a defined set of workflow states, each of which carries a specific permission scope.

Building workflow-contextual access policy requires understanding how work actually flows through the organization — not how the organizational chart says it flows, but how practitioners actually perform their tasks. A nurse accessing a patient record from a clinical workstation while in the patient's room is in a different workflow context than a billing coordinator accessing the same record from an administrative workstation for claims processing. The same data is involved, but the workflow contexts are distinct, and the permission scopes appropriate to each are different.

For managers, the workflow-contextual access policy is where zero trust architecture intersects most directly with operational process design. The conversations required to define workflow contexts — with clinical staff, administrative staff, IT operations teams, and application owners — are not primarily technical conversations. They are process conversations that happen to have security implications. Security leaders who treat this work purely as a technical configuration task will produce policies that do not align with real operational workflows and will face user resistance and exception requests that undermine the least privilege posture they are trying to build.

2.2.3 Privilege Creep: How Access Accumulates and Why It Must Be Reclaimed

Privilege creep is the gradual accumulation of access rights that occurs when permissions are granted to meet immediate operational needs but are never systematically reviewed and revoked as those needs change. It is not a sign of malicious intent or negligent administration — it is the natural outcome of operating in a dynamic environment where the pace of operational change outstrips the pace of access governance review.

The risk profile of an environment affected by privilege creep is difficult to accurately assess because the access picture has diverged from any documented policy. When an incident occurs, the forensic investigation must reconstruct what access each identity actually held — a process that consumes significant time and may not produce a complete picture if access grants were not logged comprehensively. When a user leaves the organization, offboarding processes may deprovision their primary credentials while leaving behind service account access, API tokens, and group membership entitlements that were accumulated across multiple systems over the years.

Reclaiming accumulated privilege requires a combination of technical tooling and organizational discipline. Access reviews — structured periodic audits in which application owners and managers certify whether each access grant is still operationally necessary — are the primary mechanism for surfacing and revoking orphaned or unnecessary permissions. These reviews are operationally demanding, and organizations that treat them as compliance paperwork rather than genuine risk management exercises tend to rubber-stamp existing access rather than critically evaluate it. Designing review workflows that surface real risk rather than simply generating signed certifications is an organizational design challenge as much as a technical one.

2.3 Assume Breach: Designing Systems That Expect to Be Compromised

Assume breach is the most operationally disruptive of the three zero trust principles because it requires a fundamental reorientation of the security posture. Instead of designing systems to prevent breaches as the primary goal and treating detection and response as secondary concerns that activate only after prevention fails, assume that a breach treats the attacker's eventual success as a design input. The architecture is designed not just to resist compromise but to contain its consequences, detect it as early as possible, and recover from it with minimal operational disruption.

This reorientation has immediate implications for how security investments are allocated and how security programs are measured. Prevention-focused security invests heavily in keeping attackers out. Assume breach posture invests equally in knowing when they are in and limiting what they can do once there. An organization that measures its security program only by perimeter effectiveness, blocked intrusion attempts, quarantined phishing emails, and patched vulnerabilities, will consistently under-invest in the detection and containment capabilities that assume a breach is inevitable.

Diagram 2.3 – Assume Breach Design Philosophy: Containment and Detection Architecture

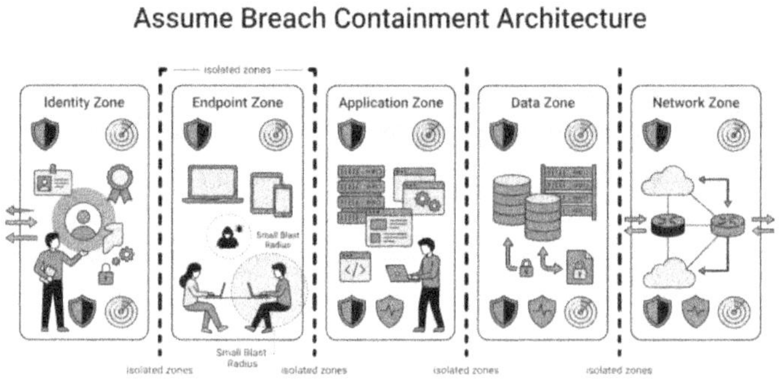

Assume Breach Containment Architecture

2.3.1 Blast Radius Minimization as an Architectural Goal

Blast radius is a measure of the damage an attacker can cause once they have obtained an initial foothold in the environment. In a flat network with broad implicit trust, the blast radius of a single compromised credential can encompass the entire enterprise: every system reachable from the internal network, every data store accessible to the compromised identity, and every lateral movement pathway available to an authenticated session. Blast radius minimization is the architectural discipline of reducing that scope through design: limiting how far any single compromised component can reach, how many resources any single identity can access, and how much damage any single session can enable.

The primary mechanisms for minimizing blast radius are network segmentation and least-privilege access — the same disciplines addressed in the previous section and in the chapters that follow. Segmentation limits lateral movement by requiring authenticated sessions to encounter policy enforcement at each segment boundary rather than traversing the internal network freely. Least-privilege limits the value of any single compromised credential by ensuring it does not grant the standing access needed to reach high-value targets without additional verification.

From a manager's perspective, blast radius is a useful frame for prioritizing security investments. Rather than asking "which assets are most valuable?" and investing to protect them individually, blast radius thinking asks "which architectural changes would most significantly reduce the damage any single compromise could cause?" That framing often produces different investment priorities that focus on eliminating flat network topology and reducing standing privilege rather than hardening individual targets.

2.3.2 Detection-First Thinking and Its Operational Implications

Detection-first thinking is the operational expression of assume breach: the security program is designed around the expectation that attackers will sometimes succeed in gaining a foothold, and the primary measure of program effectiveness is how quickly and reliably that foothold is detected and contained. Detection is not a secondary function that activates after prevention fails; it is a co-equal design priority alongside prevention, and in some contexts, the more operationally significant one.

Detection-first thinking changes how telemetry infrastructure is designed. An organization operating under a prevention-first philosophy typically collects logs on what was blocked, what was allowed, and what triggered an alert at the firewall. An organization operating under a detection-first philosophy instruments the interior of the environment as comprehensively as the perimeter: authentication events across every system, data access patterns across every data store, and lateral communication between every workload. The goal is not to generate more alerts — it is to ensure that anomalous behavior anywhere in the environment is visible to the security operations function, regardless of where it occurs.

Detection-first thinking also changes how incident response is designed. Rather than structuring response playbooks around breach notifications, the activities that occur after a breach is confirmed. A detection-first organization builds playbooks around investigation triggers: the specific behavioral signals that initiate a containment-and-investigation workflow before the breach is confirmed. This approach compresses the time between initial compromise and active response, which directly reduces dwell time and its associated costs.

2.4 Principles in Practice: Avoiding Zero Trust Theater

Zero trust theater is the condition in which an organization has adopted the vocabulary of zero trust — deploying multi-factor authentication, publishing a zero trust policy, certifying compliance against a framework — without implementing the architectural and operational changes that the principles actually require. It is not a problem unique to zero trust; most significant security paradigm shifts produce a period during which the terminology spreads faster than genuine implementation. But zero trust theater is particularly costly because it creates organizational confidence that the problem has been addressed while leaving the underlying vulnerabilities largely intact.

Identifying zero trust theater requires looking beyond certifications and product deployments to the underlying access model. The diagnostic questions are straightforward: Can an authenticated user traverse the internal network without encountering additional policy enforcement? Do service accounts hold standing access to resources beyond what current operational tasks require? Would a compromised credential from a low-privilege account provide a meaningful pathway to high-value targets? If the answers to these questions reveal that broad implicit trust relationships persist beneath a zero trust label, the implementation has not yet honored the principles.

Diagram 2.4 – Zero Trust Theater vs. Genuine Zero Trust Implementation

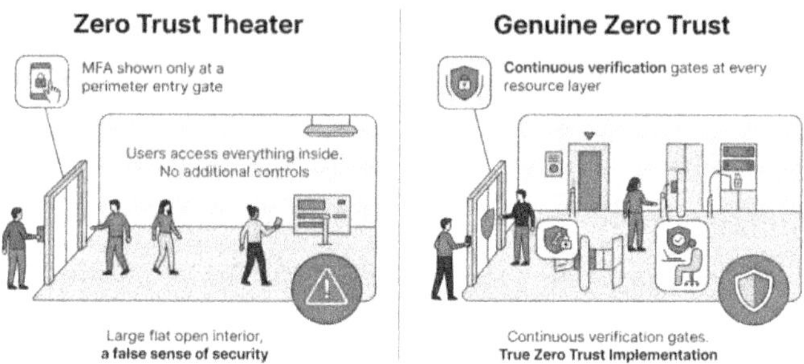

2.4.1 What Genuine Compliance with the Principles Looks Like

Genuine compliance with zero trust principles is not visible in a deployment checklist — it is visible in the access model the deployment produces. An organization genuinely implementing explicit verification has an access architecture in which every resource access event is evaluated against a current risk assessment, not simply validated against a session token issued at login. An organization genuinely implementing least privilege has a provisioning model in which standing access is the exception rather than the default, and in which access reviews surface and revoke unnecessary permissions rather than certifying them. An organization genuinely implementing assume breach has detection instrumentation inside its environment that is as comprehensive as the detection at its perimeter.

The operational indicators of genuine compliance are also observable in how the organization responds to anomalous events. A genuinely compliant organization has playbooks that define specific behavioral triggers for initiating investigation — not just alerts for confirmed malicious activity. It has containment mechanisms that can isolate a compromised session without taking down the broader environment. It has forensic capabilities that can reconstruct the full scope of a compromise using the telemetry collected by its detection infrastructure. These capabilities are the operational evidence that assume breach posture has been implemented rather than declared.

2.4.2 Evaluating Vendor Claims Against the Core Framework

The zero trust label appears across an enormous range of products and services, most of which address some dimension of the principles while leaving others unaddressed. A network access control platform that enforces identity-based segmentation addresses elements of explicit verification and least-privilege access but if it does not integrate device health signals, it is not performing the full signal aggregation that explicit verification requires. An identity provider that offers continuous access evaluation is addressing continuous authentication — but if the access decisions it produces are not scoped to least privilege, the verification is happening without the privilege discipline that makes it operationally effective.

Evaluating vendors' claims against the core framework requires mapping each vendor's capabilities to the principles and identifying gaps. No single vendor addresses all three principles comprehensively zero trust architecture is inherently multi-vendor and requires integration across identity, device management, network enforcement, and data protection domains. Vendors who claim to deliver complete zero trust in a single platform are either defining "complete" narrowly or are misrepresenting the scope of the framework. The manager who understands the principles can identify which of these is occurring and evaluate accordingly.

The evaluation process is most productive when it begins with the principal's requirements rather than the product's capabilities. Define what explicit verification requires in your specific environment, which signal sources, which policy enforcement points, and which integration dependencies. Then evaluate which products address those requirements and how they integrate. This approach produces an architecture designed around the principles, rather than a collection of products that happen to use zero trust terminology.

2.4.3 Principles as an Ongoing Standard, Not a One-Time Certification

One of the most persistent misconceptions about zero trust is that it is a project with a completion state — a set of controls that, once deployed, satisfies the requirements and allows the organization to move on. The principles do not support that reading. Explicit verification requires continuous operation of the verification infrastructure. Least privilege requires continuous maintenance of the access policy as roles, tasks, and the environment change. Assume breach requires continuous operation of detection and response capabilities that evolve as attacker techniques evolve. Zero trust is not a destination but an ongoing operational discipline.

The implications for program governance are significant. A zero trust initiative that is structured as a project — with a defined scope, a completion date, and a budget that ends when the deliverables are produced — will produce a deployment that begins decaying the moment the project closes. Within months, access rights will begin to accumulate outside the least-privilege scope, detection instrumentation will fall behind new deployment patterns, and verification policies will cease to reflect the current risk landscape. The organization will have purchased the initial deployment of zero trust without funding the operational discipline required to sustain it.

Responsible by design means structuring zero trust as an operational function rather than a project — with a defined team, a recurring budget, a governance model that reviews and updates policies on a regular cadence, and success metrics that evaluate ongoing risk reduction rather than deployment completeness. The chapters ahead address how that operational structure is built and how it sustains itself across leadership transitions, budget cycles, and evolving threat conditions.

2.5 Manager's Checklist

Test your current explicit verification posture: determine whether your authentication infrastructure evaluates device health, location, and behavioral signals, or only identity credentials, before granting access to resources.

Identify your three highest-privilege standing accounts: assess whether those accounts require just-in-time elevation or currently hold persistent administrative access that could be exploited if the credentials are compromised.

Review your last access certification: evaluate whether the process surfaced and revoked unnecessary permissions, or primarily generated signed certifications for existing access grants.

Map your detection instrumentation: determine what percentage of your internal environment generates telemetry that is actively monitored by your security operations function.

Evaluate your incident response playbooks: confirm that you have defined behavioral triggers that initiate investigation before breach is confirmed, not only after it is detected.

Apply the zero trust theater diagnostic: identify which internal resources can be accessed via an authenticated session without encountering additional policy enforcement points.

Structure your next vendor evaluation against the three principles: map each product's capabilities to explicit verification, least privilege, and assume breach — then identify the gaps your architecture must address.

Assess your program governance model: determine whether zero trust is funded and staffed as an ongoing operational function or as a time-bounded project with a completion date.

2.6 What This Chapter Establishes

The three principles of zero trust, explicit verification, least-privilege access, and assume breach, are not independent concepts that can be selectively implemented. They are interlocking design disciplines that reinforce one another. Explicit verification without least privilege yields continuous authentication, but it still grants too much access when it succeeds. Least privilege without an "assume-breach" posture yields a well-scoped access model that remains under-instrumented for detection. Assume breach without explicit verification, and least privilege produces detection investment in an environment where the blast radius of any compromise remains catastrophically large.

The operational integrity of a zero trust program depends on honoring all three principles simultaneously and maintaining that discipline over time. Zero trust theater — the adoption of the vocabulary without the architectural substance — produces a false sense of security that may be more dangerous than the perimeter model it replaced, because at least the perimeter model was honest about its limitations.

The framework these principles provide is durable because it is technology-agnostic. Products, platforms, and threat techniques will all evolve. The principles will not. A manager who can evaluate any architectural proposal or vendor claims against explicit verification, least privilege, and the assumption of breach has a stable compass for navigating a technology landscape that will continue to change throughout the program's lifetime. The next chapter translates these principles into the specific components and structural decisions that make them operational in a real enterprise environment.

Diagram 2.5 – Three Principles Integration Map

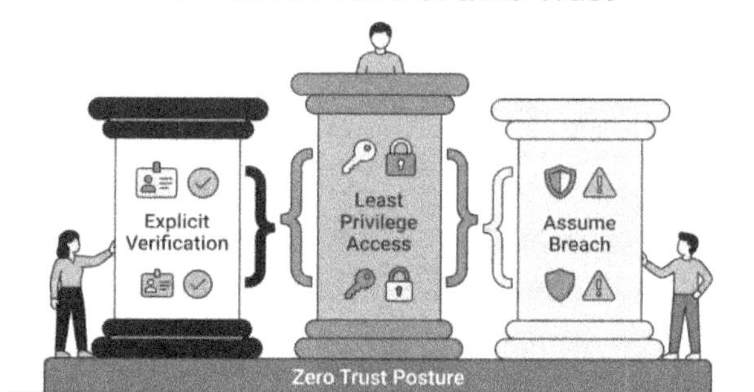

3 Assembling the Machine: Components, Control Planes, and Policy Engines

A security architect presents a zero trust deployment plan to the infrastructure committee. The slide deck shows a before-and-after network diagram: the old model with a perimeter firewall and flat internal network, the new model with a policy engine at the center and enforcement points distributed across cloud, on-premises, and remote environments. A committee member asks a direct question: when a user in Singapore tries to open a customer database on their laptop at seven in the morning, exactly what happens between the moment they click and the moment they see the data or don't? The architect pauses. The deployment plan describes the components but has not mapped the decision flow. The committee members, who need to fund and defend this architecture to their board, do not yet have the conceptual model to evaluate whether the proposed solution will actually work.

That gap between the abstract principles of zero trust and the concrete mechanical reality of how access decisions are made, enforced, and logged is what this chapter closes. Zero trust is not a philosophy that operates itself. It is an engineered system composed of specific functional components that must be understood individually and in their interactions before any deployment can be evaluated, funded, or governed with confidence. The three foundational components, the Policy Engine, the Policy Administrator, and the Policy Enforcement Point, are the machinery through which the principles become operational. Understanding how they work, how they fail, and how they must be integrated into existing infrastructure is the prerequisite for every architectural decision the chapters ahead will require.

For a manager, the component model of zero trust architecture matters for reasons that go well beyond technical literacy. Procurement decisions are made at the component level. Purchasing a product without understanding which component it addresses and which adjacent components it requires creates integration gaps that no single vendor will acknowledge. Staffing decisions depend on understanding which components require ongoing human expertise versus which can operate with higher automation. Failure mode planning requires identifying which components are single points of failure and which have built-in resilience. The organizational clarity to make these decisions correctly comes from a working model of architecture as an engineered system, not just a security philosophy.

The workflow-level impact of each component is also significant. The Policy Engine determines which organizational functions are involved in defining access policy — security teams, compliance officers, application owners, and business unit leaders, all contribute to the trust algorithms it executes. The Policy Administrator determines how session management is operationalized — which teams monitor broker activity, how session anomalies are escalated, and what the response workflow looks like when a session is terminated for policy violation. The Policy Enforcement Point determines which infrastructure teams are responsible for enforcement consistency and which teams are accountable when enforcement gaps allow unauthorized access. Operational clarity at this level requires understanding the component model before deployment begins.

3.1 The Policy Engine: Where Access Decisions Are Actually Made

The Policy Engine is the core decision-making component of a zero trust architecture. When an access request arrives from a user, a device, an application, or an automated process the Policy Engine is the system that evaluates that request against defined policy and produces one of three outcomes: access granted, access denied, or access granted with conditions such as step-up authentication or reduced session privileges. Every other component in the architecture depends on the Policy Engine's decision; none of them makes access determinations independently.

The quality of the Policy Engine's decisions depends on two factors: the completeness and accuracy of the data it evaluates, and the precision of the policy it evaluates that data against. A Policy Engine receiving incomplete signal data — missing device health information, stale behavioral baseline data, or identity signals that have not been refreshed since a previous session — will make access decisions based on a partial picture of the actual risk. A Policy Engine operating on imprecise or outdated policy rules that have not been updated to reflect current role structures, data sensitivity classifications, or threat intelligence will make decisions that are internally consistent but operationally incorrect.

Diagram 3.1 – Policy Engine Decision Flow

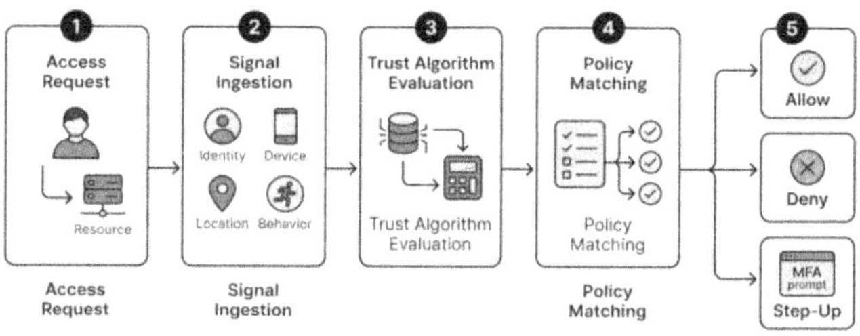

3.1.1 Trust Algorithms and the Data Inputs That Feed Them

A trust algorithm is the computational logic that the Policy Engine applies to its input signals to produce a trust score or trust classification for a given access request. The algorithm is not a simple lookup — it does not merely check whether the requesting identity is in an approved list. It evaluates a weighted combination of signals, each of which contributes to the overall assessment of whether this request, from this identity, on this device, in this context, should be trusted with access to this resource at this moment.

The input signals that feed a trust algorithm fall into several categories. Identity signals include the authentication method used, the identity provider that validated the credential, the freshness of the last authentication event, and any flags or anomalies associated with the identity in recent history. Device signals include device compliance status from the mobile device management or endpoint detection platform, operating system currency, the presence and status of required security software, and any recent behavioral anomalies flagged by endpoint analytics. Network signals include the originating IP address and the risk profile of the network from which the request is arriving — corporate LAN, known remote-access IP range, anonymous proxy, or Tor exit node — each carries different risk implications. Behavioral signals include how this access request compares to the established pattern for this identity — the time of day, the sequence of prior requests in the current session, and the sensitivity of the requested resource relative to the user's established access history.

For managers, the trust algorithm is the point where the organization's risk appetite becomes a concrete technical specification. High-sensitivity resources require higher trust score thresholds before access is granted. Deviations from baseline behavioral patterns trigger lower trust scores that may route the request to a conditional access path rather than a direct grant. The organization's tolerance for false positives — legitimate users denied access because their context looks unusual — must be balanced against its tolerance for false negatives — unauthorized access granted because the attacker's behavior looks sufficiently similar to the legitimate user's profile. These are risk management decisions, not technical configuration decisions, and they require leadership input to be made correctly.

The data quality requirements for trust algorithm inputs are significant and often underestimated. A device compliance signal that is twelve hours stale may not reflect a device that was compromised two hours ago. A behavioral baseline built on thirty days of data may not accurately represent a user who has recently changed roles. An identity signal from an identity provider that has not ingested recent threat intelligence may not flag a credential that appears on a known breach list. Maintaining the data freshness and accuracy that trust algorithms require is an operational responsibility that must be assigned, resourced, and monitored.

3.1.2 Automated Decision-Making vs. Human-in-the-Loop Escalation

The vast majority of access decisions in an enterprise environment must be made automatically and in real time. A workforce of several thousand users generating dozens of access requests per hour cannot route every request through a human review process without introducing latency that makes the architecture operationally unacceptable. Automated decision-making is therefore not a design option in a zero trust policy engine — it is a requirement. The policy must be defined precisely enough that the trust algorithm can evaluate it without human interpretation for the great majority of access events.

Human-in-the-loop escalation is appropriate for access requests that fall outside the automated decision space: requests where the trust score is ambiguous, where resource sensitivity is high enough that an automated grant carries unacceptable risk, or where the combination of signals does not align with the patterns for which automated policy has been defined. In these cases, the Policy Engine routes the request to a human reviewer with the relevant signal data and a recommended decision, rather than making a definitive grant-or-deny determination autonomously. The human reviewer applies judgment to the ambiguous case, and that judgment is logged alongside the automated decisions it supplements.

Designing the boundary between automated decisions and human-in-the-loop escalation is a policy governance question. If the escalation threshold is set too low — routing too many requests to human review — the review queue becomes unmanageable, and reviewers begin approving requests without genuine evaluation, which defeats the purpose of the escalation mechanism. Suppose the threshold is set too high — automating decisions involving significant uncertainty — the trust algorithm makes consequential decisions based on insufficient confidence. At the same time, the human judgment that should supplement it is absent. Calibrating this boundary requires operational data from actual access patterns and iterative adjustment as those patterns evolve.

3.1.3 Logging Every Decision: Auditability as a First-Class Requirement

Every decision the Policy Engine makes — grant, deny, or conditional — must be logged with sufficient detail to reconstruct the complete context of that decision. The log must capture the identity of the requesting party, the resource requested, the signals evaluated, the trust score produced, the policy applied, and the decision outcome. This logging requirement is not optional, and it is not primarily a compliance artifact — it is an operational necessity for three distinct functions.

The first function is forensic reconstruction. When an incident occurs, the investigation depends on tracing the access events that enabled the attacker's activity. If Policy Engine decisions are not logged with complete context, the investigation cannot determine what access was granted, on what basis, and whether the trust algorithm was operating correctly or was producing decisions based on corrupted or manipulated signal data. Incomplete logs lead to incomplete investigations, which lead to incomplete remediation and recurring exposure.

The second function is policy tuning. The operational data generated by Policy Engine decisions is the primary input for determining whether policy definitions produce the intended outcomes. A high rate of false positives — legitimate users being denied access or challenged for step-up authentication when the context is actually low-risk — indicates that policy definitions are too restrictive for normal operational patterns. A high rate of exceptions — human reviewers overriding automated deny decisions — indicates that automated policy has not kept pace with legitimate workflow patterns. Both patterns are only visible through comprehensive decision logging.

The third function is accountability. In regulated environments, access to sensitive data must be demonstrably governed by documented policy. The audit trail produced by

Policy Engine decision logging is the evidence that access governance is operating as designed. It demonstrates to auditors, regulators, and governance bodies that access decisions are not being made arbitrarily but are being evaluated against defined policy and logged for review. Auditability as a first-class requirement means designing logging infrastructure to evidence standards, not as an afterthought.

3.2 Policy Administrators and Enforcement Points: Translating Decisions into Action

The Policy Engine produces decisions. The Policy Administrator acts on them. The Policy Enforcement Point executes them. This sequence — decision, action, execution — is the operational chain through which a zero trust architecture moves from abstract policy to concrete access control, and each link in the chain must function correctly for the architecture to deliver on its principles. Understanding each component's distinct role prevents design errors that occur when its functions are conflated or when integration gaps between them remain unresolved.

The Policy Administrator is the component that translates a Policy Engine decision into the specific commands that configure access for a given session. When the Policy Engine grants access to a resource, the Policy Administrator issues the session token, configures the enforcement point to allow the traffic, and monitors the session for signals that should trigger re-evaluation. When the Policy Engine denies access, the Policy Administrator ensures that the denial is enforced at the appropriate enforcement point and logs the denial event. When the session ends — through expiration, user logout, or anomaly detection — the Policy Administrator revokes the session credentials and instructs the enforcement point to terminate the connection.

Diagram 3.2 – Policy Administrator and Enforcement Point Interaction

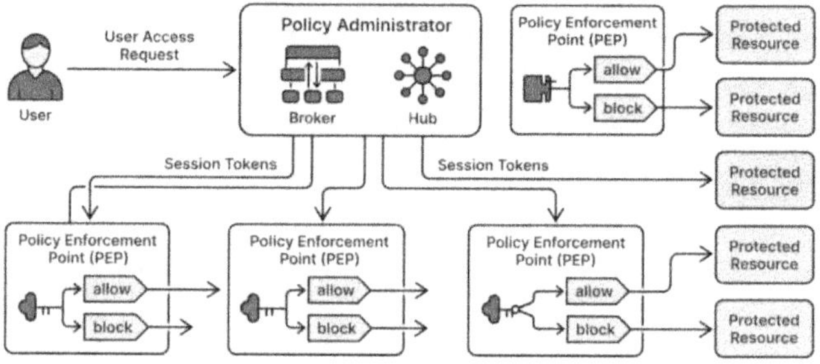

3.2.1 Session Brokering and Token Issuance

Session brokering is the operational process by which the Policy Administrator establishes and manages a specific access session following a Policy Engine grant decision. The broker creates a time-limited, scoped credential — a session token — that allows the requesting identity to communicate with the target resource for the duration of the session, within the privilege scope determined by the Policy Engine to be appropriate. The session token is not a long-lived credential; it expires, it carries the specific permissions defined for this session, and it is issued fresh for each new access event rather than reused across multiple sessions.

The session token model is central to how zero trust architecture operationalizes least privilege at the session level. A standing credential that carries the full set of permissions associated with an identity is replaced by session-specific tokens that carry only the permissions required for the current task. An attacker who intercepts a session token does not obtain a persistent credential — they obtain a time-limited, narrowly scoped credential that expires and cannot be extended without re-triggering the Policy Engine evaluation process. The value of a stolen token is therefore bounded by the duration and scope of the session it was issued for, rather than by the full access rights of the compromised identity.

Token issuance is also the mechanism through which conditional access decisions are implemented. When the Policy Engine grants access with conditions — step-up authentication, read-only scope, session monitoring elevation — the Policy Administrator incorporates those conditions into the token it issues. The enforcement point that validates the token enforces the conditions it contains, ensuring they are operationally binding rather than advisory. This integration between token issuance and enforcement point validation is a critical architectural dependency that must be designed explicitly rather than assumed.

3.2.2 Distributed Enforcement Across Hybrid Environments

Policy Enforcement Points are the components that gate access to resources — they are the mechanisms through which the Policy Administrator's decisions are translated into allow or deny actions at the network or application layer. In a traditional perimeter model, enforcement is concentrated at a small number of points: the firewall, the VPN gateway, the proxy server. In a zero trust architecture, enforcement is distributed — there are enforcement points at every resource boundary, across every environment where the enterprise operates.

In a hybrid enterprise environment that spans on-premises data centers, public cloud infrastructure, SaaS applications, and remote endpoints, distributed enforcement is a significant operational challenge. Each environment has its own native enforcement mechanisms — cloud providers offer identity-aware proxies and service mesh integrations; SaaS platforms offer SSO integrations and API gateway controls; on-premises environments offer network access control systems and application gateway proxies. Coordinating consistent policy enforcement across this diversity of mechanisms requires a Policy Administrator capable of communicating with each environment's enforcement infrastructure and a Policy Engine whose decisions can be translated into the specific control language each enforcement mechanism understands.

For managers, distributed enforcement has direct implications for staffing and vendor management. Each enforcement point type requires operational expertise — someone who understands how the cloud provider's enforcement mechanism works, how the SaaS platform's SSO integration is configured, and how the on-premises access control system is maintained. As the number of enforcement point types grows with the complexity of the enterprise environment, the expertise required to maintain consistent enforcement grows with it. This operational complexity is a real cost of the zero trust architecture that must be accounted for in staffing plans and vendor support agreements.

3.2.3 Consistency Challenges When Enforcement Points Multiply

Enforcement consistency is the property that the same access request, evaluated by the same Policy Engine, produces the same enforcement outcome regardless of which enforcement point handles it. In a simple environment with a single enforcement mechanism, consistency is relatively easy to maintain. In a complex hybrid environment with dozens of enforcement points across multiple cloud providers, SaaS applications, and on-premises systems, consistency requires systematic attention to a set of challenges that tend to compound over time.

The first challenge is the fidelity of policy translation. Each enforcement point type speaks its own policy language: the control syntax for a cloud provider's identity-aware proxy differs from that of an on-premises network access control system, which differs from the API used by a SaaS platform to enforce SSO policy. When the Policy Administrator translates a Policy Engine decision into enforcement commands, each translation must produce an outcome functionally equivalent to the intended policy — not an approximation, not a best-effort match, but an accurate representation of what the Policy Engine decided. Translation errors produce enforcement gaps: access that should be denied is permitted, or access that should be permitted is denied, because the enforcement point received an imprecise instruction.

The second challenge is enforcement point drift. Enforcement points are not static — they receive software updates, configuration changes, and integration modifications as the enterprise environment evolves. Each change has the potential to introduce an enforcement inconsistency if it is not validated against the policies the enforcement point is responsible for implementing. Configuration drift — the gradual divergence between the intended policy and the actual enforcement point configuration — is a persistent operational risk that requires automated validation tooling and regular audit processes to detect and correct.

3.3 Control Plane vs. Data Plane: Keeping Orchestration Separate from Traffic

The distinction between the control plane and the data plane is fundamental to the architecture of any zero trust deployment. It has direct implications for both security properties and operational resilience. The control plane is the network and infrastructure through which Policy Engine decisions, Policy Administrator commands, and enforcement point configurations are communicated. The data plane is the network through which the actual resource traffic — the user's query to the database, the application's call to the API, the file transfer that the enforcement point has just approved — flows. These two types of traffic must be kept separate.

The security rationale for separation is straightforward: if control plane traffic and data plane traffic share the same network path, an attacker who has compromised a data plane session has potential access to the channels through which enforcement decisions are communicated. Intercepting or manipulating control plane traffic could allow an attacker to observe access decisions, understand policy rules, or, in the most severe cases, inject false commands into the Policy Administrator's communication with enforcement points. Separation ensures that compromising a data session does not provide access to the architecture's governing logic.

Diagram 3.3 – Control Plane and Data Plane Separation in Zero Trust Architecture

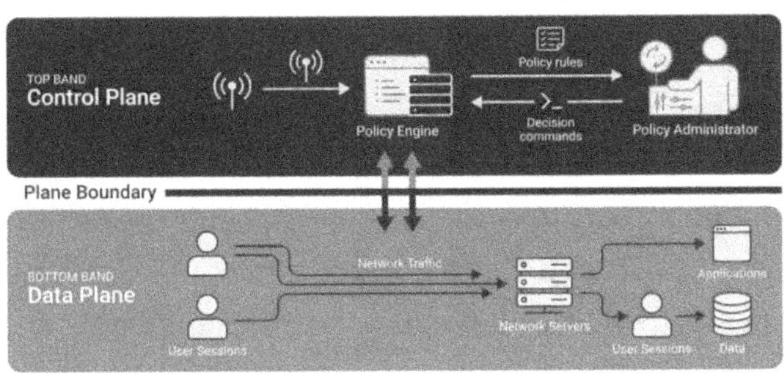

3.3.1 Why Separation Matters for Both Security and Scalability

The scalability rationale for separating the control plane and data plane is equally important but less frequently articulated. When control and data traffic share the same infrastructure, scaling the architecture to handle increased data traffic loads requires scaling the control infrastructure as well — even if the volume of access decisions has not increased. Separating the planes allows each to be scaled independently according to the demand it actually faces. High-throughput data environments can scale data-plane infrastructure without a proportional increase in control-plane capacity, and vice versa.

Separation also simplifies the security properties that each plane must provide. The control plane handles relatively low-volume, high-sensitivity traffic — policy decisions, session tokens, enforcement commands — that require strong confidentiality and integrity protection but do not require high throughput. The data plane handles high-volume, variable-sensitivity traffic that requires throughput optimization and latency minimization. Designing each plane for its specific requirements produces a more efficient architecture than attempting to satisfy both sets of requirements with a single network path.

For managers, the distinction between the control plane and data plane is relevant to procurement and vendor evaluation. Zero trust products that bundle control and data-plane functions into a single network path may be simpler to deploy initially but harder to scale and to secure as the environment grows. Products that respect the architectural separation from the beginning — with dedicated control-plane infrastructure that is isolated from data-plane traffic — are better positioned to maintain security properties at enterprise scale and over the long duration that a zero trust program requires.

3.3.2 Failure Modes and Resilience Design

Understanding how a zero trust architecture fails — not just how it functions when operating correctly — is essential for designing for resilience and operating the architecture responsibly. The control plane is the most sensitive failure point. If the Policy Engine becomes unavailable, access decisions cannot be made, and access to resources is blocked until the engine is restored. This is the correct default behavior — fail secure, not fail open — but it requires the control plane infrastructure to meet availability requirements appropriate to the operational criticality of the resources it governs.

For mission-critical resources, control-plane availability is a business-continuity requirement, not just a security requirement. If the Policy Engine that governs access to a critical patient care system becomes unavailable, clinical staff cannot access the records they need to provide care. If the engine that governs access to financial trading systems fails, trading operations stop. High-availability design for control-plane infrastructure — redundant components, geographic distribution, and defined failover procedures — must be scoped to the criticality of the resources being governed, not to the engineering team's preference for simplicity.

The data plane presents a different set of failure modes. Enforcement point failures — whether due to software errors, configuration problems, or capacity overload — can produce either fail-secure outcomes (traffic is blocked) or fail-open outcomes (traffic is permitted without policy evaluation). Fail-open outcomes are particularly dangerous in a zero trust architecture because they create gaps in the enforcement fabric through which unauthorized access can flow without generating access decision logs that the audit trail depends on. Designing enforcement points to fail securely, with explicit alerting when a failure occurs, is a non-negotiable requirement for maintaining the integrity of the zero trust enforcement model.

Resilience design for zero trust architecture requires modeling each component's failure and defining the acceptable response. Some failures can be tolerated with degraded functionality. A behavioral analytics feed that goes offline reduces the richness of trust algorithm inputs, but does not prevent access decisions from being made. Others cannot be tolerated at all. A Policy Administrator failure that prevents session tokens from being issued or revoked must trigger immediate alerting and escalated response. Building the failure mode map and the corresponding resilience design is an architectural task that must be completed before deployment, not after the first significant incident.

3.4 Reference Architectures: Federal, Healthcare, and Enterprise Patterns

Abstract component models become actionable when grounded in reference architectures that reflect real deployment environments. The federal government's approach to zero trust — driven by executive mandate and implemented across agencies with diverse technology estates and security requirements — provides the most extensively documented reference available. Healthcare and commercial enterprise patterns demonstrate how the same component model adapts to varying regulatory contexts, risk profiles, and organizational constraints. Understanding these reference architectures allows managers to locate their own deployment within a set of known patterns rather than treating every architectural decision as a novel problem.

Diagram 3.4 – NIST SP 800-207 Component Mapping

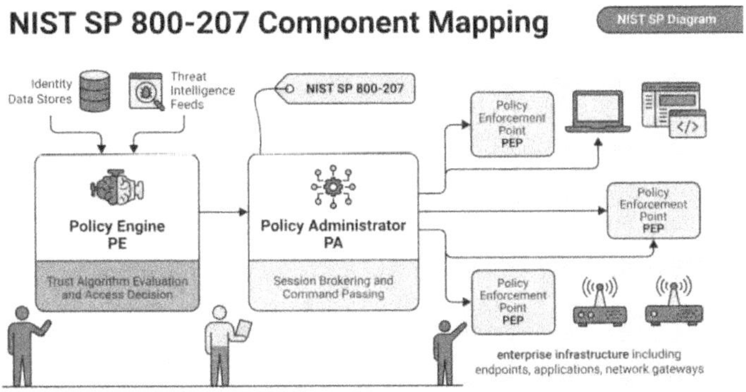

3.4.1 NIST SP 800-207 as a Deployment Blueprint

NIST Special Publication 800-207 is the most widely adopted reference framework for zero trust architecture in the United States and has influenced the development of zero trust standards globally. The publication defines the core components — Policy Engine, Policy Administrator, and Policy Enforcement Point — and describes their logical relationships and data flows. It distinguishes between trust algorithms that evaluate access requests and the policy database that encodes the organization's access rules. It addresses the network infrastructure that connects these components and identifies the external data sources — threat intelligence feeds, identity provider signals, device compliance databases — that feed the trust algorithm.

SP 800-207's value as a blueprint is not that it specifies which products to use — it does not. Its value is that it provides component vocabulary and a vendor-neutral logical architecture, allowing organizations to evaluate any vendor's product against a stable reference without being captive to that vendor's own architecture framing. A procurement process structured around SP 800-207 components can systematically identify which components each candidate product addresses, which it does not, and which integration dependencies must be resolved through additional products or custom development.

Federal agencies implementing zero trust under executive mandate have used SP 800-207 as the primary reference architecture, supplemented by agency-specific implementation guidance from the Cybersecurity and Infrastructure Security Agency and the Office of Management and Budget. The published implementation experiences of these agencies — including the challenges they encountered around legacy system integration, identity consolidation, and enforcement point deployment — provide a practical supplement to the publication's logical architecture that commercial enterprises can draw on even though their regulatory context differs.

3.4.2 Mapping Components to Real Infrastructure Inventories

The logical component model of zero trust architecture must be mapped to the organization's actual infrastructure inventory before any deployment plan can be developed. This mapping exercise serves two purposes: it identifies which components of the zero trust architecture already exist in some form within the current environment — which products or systems could serve as the foundation for a Policy Engine, which systems are already functioning as enforcement points even if they are not labeled as such — and it identifies which components must be built or acquired.

In most enterprise environments, the mapping exercise produces a more optimistic picture than the starting assumption. Organizations that have deployed enterprise identity management systems have the foundation for the identity signal infrastructure required by the trust algorithm. Organizations that have deployed endpoint detection and response platforms have the foundation for collecting device signals. Organizations that have deployed application delivery controllers or API gateways have an enforcement point infrastructure that can be integrated into the Policy Administrator's control model. The zero trust architecture is frequently not built from scratch — it is assembled by connecting and extending existing systems, with new components added to fill the gaps identified by the mapping exercise.

The mapping exercise also identifies the integration challenges that will define the deployment's timeline and complexity. An identity-based Policy Engine cannot govern legacy applications that do not support modern authentication protocols without a proxy layer that translates their authentication model. Data stores that do not emit structured access logs cannot contribute to the trust algorithm's behavioral signal collection without additional instrumentation. Directory services that have not been consolidated following mergers and acquisitions cannot provide the unified identity foundation that the Policy Engine requires without a federation or consolidation initiative that must precede the zero trust deployment. Identifying these dependencies early — through a systematic mapping exercise rather than discovering them mid-deployment — is the difference between a phased deployment plan and a project that stalls when it encounters its first major integration obstacle.

3.4.3 Selecting Deployment Models: Agent-Based, Proxy-Based, and Hybrid

Zero trust enforcement can be implemented through several deployment models, each with different technical requirements, operational characteristics, and suitability for different environment types. Understanding the trade-offs among these models is essential for designing a deployment that matches the organization's infrastructure reality and operational constraints.

Agent-based deployment places software on each endpoint and server that enforces policy decisions locally, in communication with the central Policy Engine and Policy Administrator. The agent intercepts access requests, queries the Policy Engine for a trust evaluation, and enforces the resulting decision before the traffic reaches the target resource. This model provides fine-grained enforcement at the endpoint level and allows the agent to collect device health signals directly rather than infer them from external sources. Its operational requirements — deploying and maintaining agents on every endpoint in the environment — are significant, and legacy systems that cannot run modern endpoint software require alternative approaches.

Proxy-based deployment routes traffic through a central or distributed proxy infrastructure that performs the Policy Engine query and enforcement function in the network path rather than on the endpoint. Users and applications do not need agents — they connect to the proxy, which evaluates the request, communicates with the Policy Engine, and either proxies the traffic to the target or terminates the connection. This model is suitable for environments with diverse endpoint types, legacy systems that cannot run agents, or an architecture where centralizing enforcement simplifies consistency across endpoints. Its limitations include the latency introduced by proxy insertion into the traffic path and the availability requirements that centralized proxy infrastructure must meet.

Hybrid deployment combines elements of both models, using agents where endpoint coverage is feasible and proxies where it is not. This approach matches the real composition of most enterprise environments — a mix of managed endpoints that can run agents, unmanaged or legacy endpoints that require proxy-based enforcement, and cloud workloads that benefit from cloud-native enforcement mechanisms. Designing the hybrid model requires explicit decisions about which enforcement approach applies to each environment segment and how the Policy Administrator maintains consistent session management across enforcement mechanisms operating on different technical models.

For managers, deployment model selection is a decision that locks in significant operational commitments. An agent-based deployment requires a device management infrastructure capable of deploying and maintaining agents on a scale. A proxy-based deployment requires a proxy infrastructure capable of meeting the availability and performance requirements of the resources it governs. A hybrid deployment requires expertise in both models and an integration architecture that maintains consistency across them. The deployment model decision should be made with a clear view of the organization's infrastructure reality, operational capacity, and long-term evolution plans — not based on a vendor's preferred deployment narrative.

Diagram 3.5 – Deployment Model Comparison: Agent-Based vs. Proxy-Based vs. Hybrid

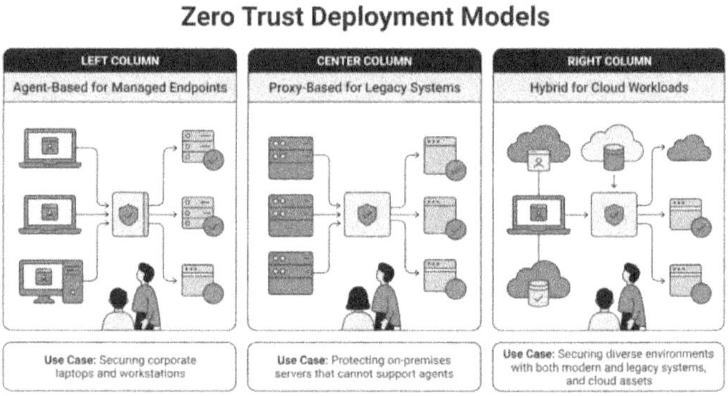

Zero Trust Deployment Models

3.5 Manager's Checklist

Map your Policy Engine candidates: identify which existing products or platforms could serve as the foundation for a Policy Engine in your environment, and assess what signal integration gaps must be addressed.

Define your auditability requirements: determine the retention period, format, and access controls for Policy Engine decision logs, and confirm that your logging infrastructure can meet those requirements at scale.

Inventory your enforcement points: catalog every environment — cloud accounts, SaaS applications, on-premises systems — where a Policy Enforcement Point must operate and identify which have native enforcement capabilities and which require proxy or agent deployment.

Assess your control plane availability requirements: identify which resources, if the Policy Engine were unavailable, would require emergency access procedures, and confirm that your high-availability design matches the criticality of those resources.

Verify your control and data plane separation: confirm that your architecture design routes Policy Engine communications through isolated infrastructure that is not shared with resource traffic, and that enforcement point failures default to deny rather than permit.

Conduct an SP 800-207 component mapping exercise: map each component in the NIST reference architecture to either an existing product in your environment or an identified gap requiring acquisition or development.

Define your deployment model for each environment segment: document whether each segment will use agent-based, proxy-based, or native cloud enforcement, and confirm that the Policy Administrator can manage session brokering consistently across all models.

Assign operational ownership for each component: identify which team is responsible for the Policy Engine, Policy Administrator, and each category of Enforcement Point, and confirm that ownership extends to ongoing policy maintenance, not just initial deployment.

3.6 What This Chapter Establishes

Zero trust architecture is an engineered system with specific functional components that must be understood individually and in their operational interactions. The Policy Engine makes access decisions. The Policy Administrator acts on them. The Policy Enforcement Point executes them. Each component has defined responsibilities, specific data dependencies, distinct failure modes, and operational ownership requirements that must be addressed before deployment begins. Treating zero trust as a philosophy rather than an engineered system yields deployments that honor the terminology while omitting the underlying mechanics.

The control plane and data plane distinction is not an abstract architectural preference — it is a security and scalability requirement that must be respected in the design of every component interaction. Enforcement consistency across distributed enforcement points is an operational discipline, not a deployment checkbox. Auditability is a first-class requirement, not a compliance afterthought. Each of these principles has a concrete operational expression that must be resourced and maintained.

The reference architectures — anchored in NIST SP 800-207 and illustrated by federal and enterprise deployment patterns — provide a stable vocabulary and logical structure for evaluating products, planning deployments, and communicating architectural decisions to non-technical stakeholders. The component mapping exercise that translates these reference architectures into the organization's actual infrastructure inventory is the critical analytical step between understanding the architecture and executing it. The chapters ahead build on this component foundation to address the specific domains — identity, network, and data — where zero trust architecture must be implemented to fulfill the principles on which it is built.

Diagram 3.6 – Chapter Summary: Zero Trust Component Architecture Overview

Complete Zero Trust Architecture

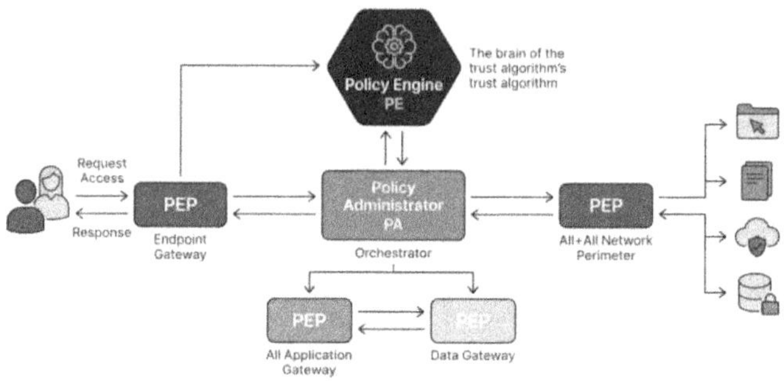

4 Identity as the New Security Perimeter: Humans, Machines, and Everything In Between

The security team at a regional healthcare network completed its annual access review on a Thursday afternoon and closed the ticket. Forty-eight hours later, an attacker moved laterally through four internal systems using credentials belonging to a nurse practitioner who had transferred to a partner clinic eight months earlier. Her account had never been deprovisioned. Her access token still worked. The firewall never flagged anything because, from the network's perspective, the session originated from a known internal address. The attack surface was not a misconfigured server or an unpatched application. It was an identity the organization had forgotten it owned.

This scenario is not unusual. It is the predictable consequence of a security model that treats identity as an afterthought — something to configure during onboarding and revisit only during compliance audits. Zero trust inverts that assumption entirely. When the network edge ceases to be the primary trust boundary, identity steps into that role, and the scope of identity that a modern enterprise must govern is vastly wider than most programs acknowledge.

Why it matters is straightforward at the mission level: the access decision is now the security event. Every time a user, a workload, or an automated process requests access to a resource, that request is a trust assertion that either the system validates or accepts on faith. Organizations that accept it on faith — because the request came from inside the network, because it created the account, because no one flagged it in the last review — have delegated their security posture to the attacker's discretion.

The workflow-level impact is equally concrete. Teams that lack a coherent identity inventory cannot enforce least-privilege policies. They cannot answer the question: who has access to what, and whether they should. When that question goes unanswered, over-permissioned accounts accumulate, orphaned credentials persist, and the attack surface grows with every hiring cycle, every acquisition, and every automation pipeline that generates a new service account. The program that cannot answer the inventory question cannot be a zero trust program in any meaningful sense.

The manager's decision that anchors this chapter is not which identity vendor to select. It is whether the organization has committed to treating identity governance as load-bearing architecture — not as a human resources workflow or a compliance checkbox, but as the primary enforcement mechanism through which access policy is operationalized. That commitment determines everything else.

4.1 The Full Roster of Identities That Zero Trust Must Account For

Most identity programs are built around human employees logging into applications. That framing is dangerously incomplete. A production enterprise environment contains at least three distinct identity categories, each with its own lifecycle, risk profile, and governance requirements. A zero trust program that governs only one of them leaves the other two as unmonitored attack surfaces.

4.1.1 Human Identity: Employees, Contractors, and Temporary Stakeholders

Human identities are the most visible and the most politically complex to govern. Full-time employees arrive with clear onboarding processes and, in theory, equally clear offboarding procedures. In practice, the offboarding side of the equation often breaks down. Accounts persist beyond end dates. Access to specific applications is never revoked because the departing employee's manager did not know which systems they had been granted access to. The identity lingers, and with it the credential.

Contractors and temporary stakeholders introduce additional complexity. They are often granted access under deadline pressure, with scope defined loosely as "what they need for the project." When the project ends, their accounts may not be tied to an HR system that triggers automatic deprovisioning. A contractor engagement that ran six months may leave an active directory account that remains valid for years afterward. Temporary stakeholders — auditors, legal counsel, implementation partners — represent a third category that many organizations handle entirely through shared credentials or ad hoc provisioning, bypassing identity governance entirely.

The manager's obligation here is to verify that every human identity category is covered by a lifecycle process that includes automated deprovisioning triggers, not just manual review. The question to put to your identity team is not "do we have an offboarding process" but "what is the longest a terminated user's account has remained active in the last twelve months, and what is the reason?"

Diagram 4.1 – Human Identity Lifecycle in a Zero Trust Environment

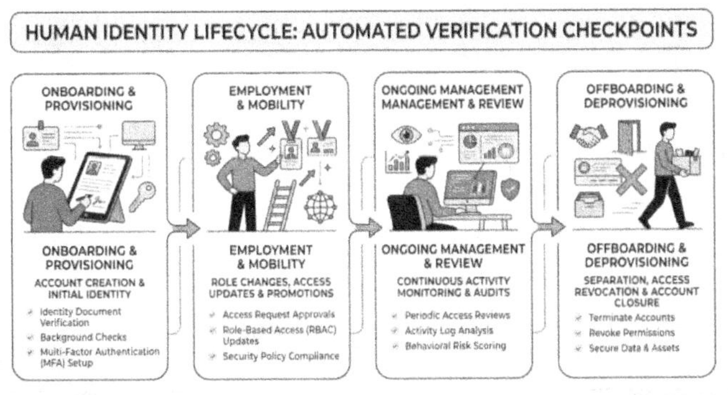

4.1.2 Machine Identity: Certificates, Tokens, and Workload Credentials

Non-human machine identities now outnumber human identities in most enterprise environments by a significant margin. Every containerized workload, every microservice, every automated pipeline, and every cloud function that communicates with another resource does so under some form of credential. Those credentials are machine identities, and they carry the same risk as human credentials — arguably more, because they often operate at higher privilege levels and without the visibility that human accounts receive.

The dominant form of machine identity is the certificate or token issued by a certificate authority or identity provider.

These credentials authenticate workloads to one another, authenticate services to external APIs, and attest to the integrity of communication channels. When they are managed well — with short lifespans, automated rotation, and cryptographic validation at each use — they function as effective trust assertions. When they are managed poorly — with long-lived certificates, manual rotation, and no central inventory — they become the technical equivalent of master keys left in unlocked drawers.

A machine identity inventory is not optional in a zero trust architecture. The organization must know which workloads hold which credentials, when those credentials expire, whether rotation is automated, and who is responsible for the infrastructure that generates them. Without that inventory, a single compromised certificate can propagate access across an environment in ways that no network control will catch, because the access is technically authenticated.

4.1.3 Service Accounts and API Keys: The Overlooked Attack Surface

Service accounts represent the identity category that generates the most risk per unit of organizational attention. They are typically created to allow one system to communicate with another — a backup application that needs database access, a monitoring tool that queries infrastructure APIs, a CI/CD pipeline that deploys to production. They are created under operational pressure, often with broad permissions granted to avoid friction. And then they are forgotten.

API keys compound the problem. Unlike service accounts, which are usually tracked at least nominally in a directory system, API keys are often generated directly by developers, stored in configuration files or environment variables, and never registered in any central inventory. A developer who leaves the organization may have generated a half-dozen API keys for personal projects or experimental pipelines. Those keys remain valid until someone explicitly revokes them, and that revocation depends on someone knowing they exist.

The zero trust posture for service accounts and API keys requires three things: a complete, continuously updated inventory; a policy that prohibits interactive logins for service accounts; and a rotation schedule that reduces the window of exposure for any single compromised credential. Operationally, this means that the team creating a new service account must register it, specify its intended scope and lifespan, and identify an owner accountable for its deprovisioning. That process cannot be optional.

Diagram 4.2 – Machine and Service Account Identity Inventory Framework

COMPREHENSIVE IDENTITY INVENTORY & GOVERNANCE FRAMEWORK

HUMAN IDENTITIES	MACHINE IDENTITIES	SERVICE ACCOUNTS & API KEYS
· STRONG AUTHENTICATION (MFA)	· CERTIFICATE MANAGEMENT (ISSUANCE & RENEWAL)	· PRINCIPLE OF LEAST PRIVILEGE
· ROLE-BASED ACCESS CONTROL (RBAC)	· SECURE SECRET & KEY STORAGE (VAULTING)	· STRICTLY SCOPED PERMISSIONS
· PRIVILEGED ACCESS MANAGEMENT (PAM)	· AUTOMATED CREDENTIAL ROTATION	· KEY ROTATION POLICIES & EXPIRY
· USER LIFECYCLE MANAGEMENT (ON/OFFBOARDING)	· ASSET INVENTORY & REGISTRATION	· USAGE LOGGING & MONITORING
· REGULAR ACCESS REVIEWS & ATTESTATION	· LEAST PRIVILEGE FOR M2M COMMUNICATION	· IDENTITY OWNER & ACCOUNTABILITY
· POLICY ENFORCEMENT & COMPLIANCE	· AUDIT TRAILS & ACTIVITY LOGGING	· REVOCATION & DEACTIVATION MECHANISMS

4.2 Strong Authentication Without Friction: Making MFA Work at Scale

Authentication is where identity governance becomes visible to the workforce. A policy that specifies strong authentication is only as effective as the user experience it delivers. When authentication is cumbersome — repeated prompts, incompatible authenticator apps, inaccessible recovery flows — users find workarounds, administrators grant exceptions, and the policy erodes in practice even as it remains intact on paper. The manager's challenge is not choosing the strongest authentication method available. It is choosing the strongest method that the organization can sustain at scale without generating a parallel track of exceptions that undermines the entire framework.

4.2.1 Phishing-Resistant Authentication Methods and When to Require Them

Standard multi-factor authentication — the push notification to a mobile app, the one-time passcode from an authenticator application — provides meaningful protection against credential stuffing and password spray attacks. It does not provide meaningful protection against real-time phishing attacks in which an adversary intercepts both the password and the MFA code during a session designed to capture them. The attacker simply relays the captured credentials to the legitimate site before the one-time code expires. The MFA event was completed successfully. The session belonged to the attacker.

Phishing-resistant authentication methods address this gap by binding the authentication event to the request's origin. FIDO2-compliant hardware security keys and passkeys are the most widely deployed examples. When a user authenticates with a FIDO2 key, the key cryptographically signs a challenge that includes the relying party's verified domain. A phishing site that proxies the authentication cannot replay the signed response against the legitimate site, because the domain binding will not match. The authentication fails at the point of attack rather than after the attacker has already established a session.

The manager's decision is which populations require phishing-resistant authentication and under what timeline. The answer is rarely "everyone immediately." A practical sequencing approach starts with the highest-risk identity categories: privileged administrators, executives, employees with access to sensitive data, and any account that can modify authentication infrastructure. Once those populations are covered, the program extends outward based on risk tier.

4.2.2 Passwordless Pathways and Their Trade-Offs

Passwordless authentication eliminates the shared secret, replacing the password with a cryptographic proof of identity that cannot be intercepted or replayed. The user experience — a biometric scan, a hardware token tap, or a platform authenticator — is typically faster and simpler than entering a password and then a secondary factor. The security posture is stronger because there is no credential to steal. The operational challenge is the transition.

Organizations that move toward passwordless authentication must account for legacy applications that still require password-based authentication, recovery scenarios for users who lose their primary authenticator, and the infrastructure required to provision and manage hardware tokens at scale. The transition period — during which some systems accept passwords and others do not — creates a hybrid authentication environment that requires careful policy coordination to prevent gaps. A user who is passwordless for their primary applications but still uses a password for a legacy system that connects to the same data store has not achieved a meaningfully stronger security posture for that data.

The path to passwordless is a multi-year initiative for most organizations, not a project that lasts a quarter. The manager should expect to run parallel authentication systems for an extended period and should define clear criteria for when a system becomes eligible for passwordless migration rather than leaving that determination to individual application owners.

4.2.3 Balancing Security with Usability in High-Volume Authentication Environments

Contact center agents authenticating dozens of times per shift, clinical staff moving between shared workstations in a hospital ward, warehouse workers sharing devices on a production floor — these are not edge cases. They are common enterprise realities that standard authentication design does not accommodate well. An authentication policy designed for knowledge workers at dedicated desks will generate significant friction in these environments, and that friction will produce workarounds: credentials written on sticky notes, shared logins for shift handoffs, and disabled authentication prompts on shared devices.

Risk-adaptive authentication provides one path through this problem. Rather than applying the same authentication requirements to every session, the policy engine evaluates the risk signal of each access request — device trust state, network location, time of day, behavioral pattern — and adjusts the authentication challenge accordingly. A clinician logging in from a managed workstation in the hospital building during their scheduled shift faces a lightweight authentication challenge. The same clinician logging in from an unmanaged device at 2:00 AM on a day they are not scheduled faces a stronger one. The security response is proportionate to the risk, and routine access remains low-friction.

Diagram 4.3 – Risk-Adaptive Authentication Decision Flow

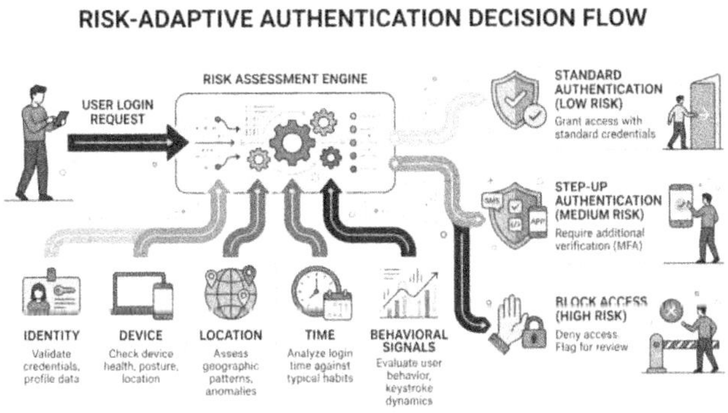

4.3 Privileged Access Management Inside a Zero Trust Framework

Privileged access is where breaches become catastrophic. An attacker who compromises a standard user account has gained a foothold. An attacker who compromises a privileged account — a domain administrator, a cloud platform owner, a database superuser — has gained the ability to move anywhere in the environment, exfiltrate at scale, and modify or destroy the controls designed to detect them. Privileged access management is therefore not a feature of identity governance but its most consequential component.

The conventional approach to privileged access assigned standing administrative accounts to IT staff, often with broad permissions that covered far more than any individual needed for their daily work. Those accounts were logged in persistently, used for both privileged tasks and routine ones, and rarely subject to the same lifecycle scrutiny as standard user accounts. The zero trust approach is fundamentally incompatible with that model.

4.3.1 Vaulting, Session Recording, and Just-in-Time Elevation

Credential vaulting addresses the most immediate problem: privileged credentials stored in people's heads or local password managers. A PAM platform vaults privileged credentials centrally, rotates them automatically after each use, and brokers access in a way that the administrator never sees the credential itself — they authenticate to the PAM platform, which authenticates to the target system on their behalf. The credential is never transmitted directly to the requesting user, which means a compromised endpoint cannot capture it.

Session recording extends this model by capturing a full audit trail of every privileged session — keystrokes, commands executed, files accessed, configuration changes made. That record serves two purposes. It creates accountability: administrators who know their sessions are recorded operate with awareness that their actions are attributable. And it creates forensic evidence: when a privileged account is suspected of misuse, the session record provides investigators with a precise timeline rather than log fragments that must be reconstructed from multiple sources.

Just-in-time elevation completes the framework by eliminating standing privilege entirely. Rather than holding a permanent administrative account, an administrator submits a request for elevated access to a specific system for a defined duration and purpose. The PAM platform evaluates the request against policy — does the requester have authorization for this type of access, is the requested scope appropriate, has a ticket been filed — and grants a time-limited credential that expires automatically. When the session ends or the time window closes, the privilege disappears. There is no persistent administrative identity to compromise.

Diagram 4.4 – Just-in-Time Privileged Access Workflow

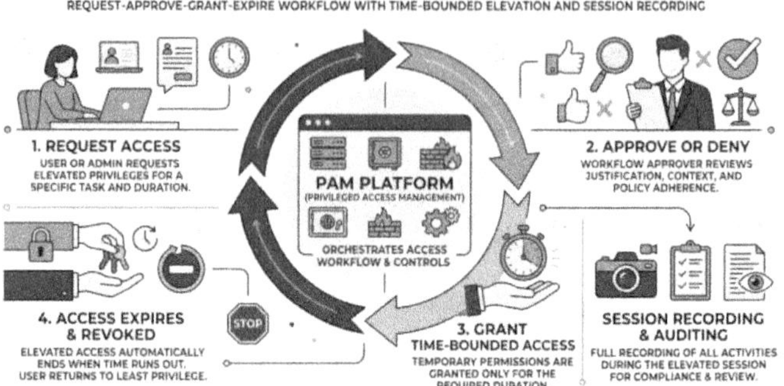

4.3.2 Breaking the Always-On Admin Account Model

Eliminating standing privileged accounts is one of the highest-return investments in a zero trust program, and one of the most organizationally difficult to execute. IT staff who have held administrative privileges for years experience the shift as a reduction in capability and autonomy. The workflow change is real: tasks that previously required clicking a menu item now require submitting a request, waiting for approval, completing the work, and closing the session. For administrators accustomed to moving quickly across systems, this cadence feels inefficient.

The response to this resistance is not to force the change without explanation but to make the operational case. Standing administrative accounts are the most targeted credentials in any environment. Attacks specifically designed to harvest administrative credentials — pass-the-hash, Kerberoasting, DCSync — are mature, widely available, and effective against standing accounts in ways they cannot be against time-limited, vaulted credentials. The friction of just-in-time elevation is a deliberate architectural property, not a flaw in the workflow. It is the mechanism through which the always-on attack surface is eliminated.

The manager's role is to set the expectation clearly and to ensure that the just-in-time workflow is calibrated to actual operational needs. Approval SLAs must be defined so that emergency access is available within minutes, not hours. Breakglass procedures must exist for scenarios where the PAM platform itself is unavailable. Emergency access credentials must be sealed, monitored, and audited immediately after use. The goal is not to prevent administrative access but to ensure that every instance of it is recorded, authorized, and time-bounded.

4.4 Federated Identity, Directory Hygiene, and the Sprawl Problem

Most organizations of any significant size carry identity stores from multiple eras. A core Active Directory domain from the early networking era. A cloud identity provider added when Microsoft 365 or Google Workspace was adopted. A separate identity store acquired when a business unit was purchased. A developer identity platform introduced by an engineering team that moved faster than the enterprise IAM roadmap. These stores overlap, conflict, and in some cases govern access to the same resources under different policies. The zero-trust architecture cannot enforce consistent policy in an environment where identity itself is fragmented.

4.4.1 Unifying Fragmented Identity Stores After Mergers and Acquisitions

Mergers and acquisitions are the most common source of identity sprawl, and they are particularly difficult to address because the business timelines of integration rarely align with the security timelines of identity consolidation. The acquiring organization wants systems connected and employees productive within weeks. A proper identity integration — assessing the acquired organization's directory hygiene, mapping trust relationships, establishing federation, eliminating duplicates, and applying consistent policy — takes months under the best circumstances.

The interim period is the danger zone. During the window when both identity stores exist independently but are loosely connected — typically through a domain trust or a federation agreement — the security posture of the combined organization is only as strong as the weaker of the two. An attacker who gains a foothold in the acquired organization's identity infrastructure can use the trust relationship to pivot into the acquirer's environment. The attack crosses the organizational boundary along a technical path designed to enable legitimate access.

The manager's guidance for M&A scenarios is to treat identity integration as a security-critical dependency, not an IT administrative task. A dedicated integration workstream with defined milestones, clear ownership, and executive visibility must exist from the day the acquisition closes. The goal is not to complete the integration before any business connectivity is established — that is operationally unrealistic

— but to ensure that interim connectivity is minimally scoped and actively monitored, and that a clear timeline exists for achieving consolidated identity governance.

Diagram 4.5 – M&A Identity Integration Phases

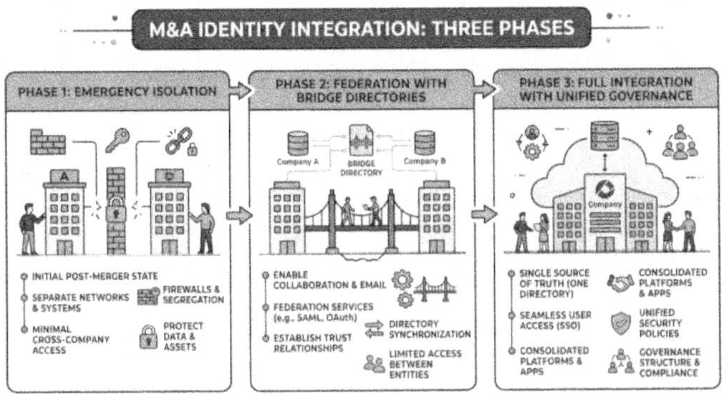

4.4.2 Lifecycle Management: Provisioning, Reviewing, and Deprovisioning at Scale

Identity lifecycle management is the operational discipline that keeps the access inventory accurate over time. It covers three processes that must work in sequence: provisioning the right access at onboarding, adjusting that access when roles change, and removing it when the relationship ends. All three processes must be automated to be reliable at scale. Manual provisioning and deprovisioning introduce delay, inconsistency, and human error at every step.

Automated provisioning connects to the HR system as the authoritative source of employment status and role. When an employee is hired into a specific job function, the identity system provisions the access appropriate to that function without requiring IT to interpret a ticket. When the employee's role changes, the system adjusts access according to a role-based policy rather than waiting for a request that may or may not be submitted. When the employee leaves, the HR system's termination event triggers automatic account suspension and a deprovisioning workflow that covers all governed systems.

Access reviews — the periodic validation that existing access remains appropriate — serve as the catch mechanism for drift that the automated process misses. A role change that occurred informally, access granted on an exception basis that was never updated in the role policy, a contractor engagement extended beyond its original scope: these are the patterns that accumulate between automated events. Access reviews surface them, but only if they are designed to surface real risk rather than generate rubber-stamp confirmations. The review should present each reviewer with the specific access being certified and require an explicit justification for continued access, not simply an approval click.

4.4.3 Governing AI Agent Identities in Automated Pipelines

AI agents represent the newest and least-governed category of identity in most enterprise environments. An AI agent — a model-backed process that takes actions autonomously in response to prompts or triggers — is neither a human identity nor a conventional service account. Still, it needs both a stable identity and scoped permissions to operate. When it calls an API, reads a file, or modifies a record, that action is attributable to it, and the permissions it holds determine what it can do.

The governance gap is significant. Many organizations deploying AI agents have not established standards for how those agents are credentialed, what access they should have, how their actions are logged, or who is accountable when an agent behaves unexpectedly. The agent exists in production, acts on behalf of users or business processes, and holds credentials that grant it access to sensitive systems — without any of the oversight structures that would be applied to a human or even a conventional service account in the same environment.

The zero trust approach to AI agent identity begins with the same principles applied to any other identity: minimum necessary permissions, short-lived credentials, comprehensive audit logging, and a defined owner who is accountable for the agent's behavior. The agent's identity should be registered in the same governance framework as all other identities. Its access should be reviewed on the same cadence. When it is decommissioned, its credentials should be revoked just as a departing employee's would be. The novelty of the technology does not exempt it from the governance requirements that all other identities are subject to.

Diagram 4.6 – AI Agent Identity Governance Model

AI AGENT IDENTITY GOVERNANCE LIFECYCLE

4.5 Manager's Checklist: Identity as Security Perimeter

Use this checklist to assess the operational readiness of your identity governance program within a zero trust framework.

Confirm that a complete identity inventory exists covering human, machine, and service account categories — not just Active Directory users.

Verify that automated deprovisioning is triggered by HR system events, not by manual IT tickets, and audit the longest active post-termination account in the last twelve months.

Establish a service account registration policy that requires owner assignment, scope documentation, and rotation schedule before any new service account is activated.

Define the privileged identity population that requires phishing-resistant authentication and set a timeline for full coverage of that population.

Confirm that your PAM platform supports just-in-time elevation, session recording, and automatic credential rotation, and verify that standing admin accounts are being eliminated according to a defined schedule.

Review the identity integration status of any business unit acquired in the last three years and confirm that an active consolidation workstream exists with defined milestones.

Identify every AI agent operating in production and confirm that each holds a registered identity, scoped credentials, an active owner, and an audit log accessible to your security operations team.

Verify that access reviews for high-risk access are designed to require explicit justification rather than passive approval, and review the exception rate from the last completed review cycle.

4.6 What This Means for Your Program

Identity governance is the load-bearing wall of a zero trust architecture. Every other control layer — network segmentation, data encryption, behavioral monitoring — depends on accurate, current, and enforced identity data to function as designed. An organization that invests in network controls while tolerating orphaned accounts, over-permissioned service accounts, and ungoverned AI agent identities has built a secure door on a building with open windows.

The path from where most organizations are — identity managed primarily as an HR support function — to where zero trust requires them to be — identity managed as the primary access enforcement mechanism — is not a technology procurement journey. It is an organizational discipline journey. The tools exist. The gap lies in process ownership, lifecycle automation, and the sustained executive attention required to close the exceptions that accumulate faster than manual review cycles can address them.

The manager who leaves this chapter with one commitment should commit to answering the inventory question: who has access to what, under what conditions, and is that access still appropriate? When that question can be answered with confidence and at any time — not just at the end of a review cycle — the identity perimeter is doing its job.

97

5 Redesigning the Network from the Inside: Microsegmentation and East-West Traffic Control

The incident began with a phishing email sent to an accounting clerk on a Tuesday morning. By Thursday evening, the attacker had reached the payroll database, the customer records archive, and two internal file servers that had not been accessed from the accounting department in any recorded period. The initial compromise took twelve minutes. The lateral movement took two days. The network had no controls between any of those systems — they all resided on the same flat internal subnet, accessible to anything that authenticated with a valid domain credential. Once the attacker had one credential, the internal network was effectively open.

Lateral movement is the defining characteristic of post-breach propagation in flat network environments. It is not a sophisticated technique. It requires no novel exploit. It requires only that the attacker move from one reachable system to another, using legitimate protocols and authenticated sessions, until they reach an asset of sufficient value. Every organization that has not implemented network segmentation is operating with an architecture that actively facilitates this movement.

Why microsegmentation matters to the manager is not primarily about the technical elegance of software-defined policy. It is about blast radius. When a breach occurs — and the architectural posture of zero trust is to assume that it will

— the question that determines organizational impact is how far the attacker can travel from the initial point of compromise. In a flat network, the answer is: everywhere the network reaches. In a segmented network, the answer is: only as far as the policy permits. That difference is measured in the scope of the forensic investigation, the number of affected records, the cost of remediation, and the regulatory consequence of the breach.

The workflow-level impact of microsegmentation is also significant for IT operations. Segment boundaries create natural observation points where traffic can be inspected, logged, and analyzed. The absence of segmentation means that east-west traffic — the communications between servers, services, and workloads within the internal network — is largely invisible. There is no chokepoint at which to examine it, no mechanism to compare it against an expected baseline. Segmentation does not just limit attacker movement. It creates the visibility architecture that enables detection of anomalous movement.

The manager's framing for this chapter is the segmentation project as an organizational initiative, not a network engineering exercise. The technical work is real and complex. But the decisions that determine whether the project succeeds — which systems belong together, which communications are legitimate, which segment boundaries represent acceptable operational disruption — are business and operational decisions that require manager-level engagement, not just engineering judgment.

5.1 Flattening the Attack Surface: Why Microsegmentation Is a Zero Trust Requirement

Microsegmentation is the practice of dividing an internal network into small, policy-controlled zones where traffic between zones is explicitly authorized rather than implicitly permitted. The name is somewhat misleading — the goal is not necessarily to create an enormous number of tiny segments, but to ensure that no system can reach another without traversing a policy enforcement point. The granularity of segmentation should be proportionate to the sensitivity of the systems involved and the operational cost of enforcing boundaries between them.

5.1.1 Mapping Communication Dependencies Before Drawing Segment Boundaries

The most consequential mistake in segmentation projects is drawing boundaries before understanding what currently crosses them. Organizations that begin with a policy framework and then attempt to enforce it frequently discover that legitimate business processes they did not know about break silently under the new rules. A batch job that runs at 2:00 AM to synchronize records between systems. A monitoring agent that calls home to a management platform using a port that the new policy does not permit. A legacy integration that connects a production application to a staging environment for reasons that were documented in a project plan from six years ago and exist nowhere else.

The dependency mapping phase is not optional and cannot be abbreviated. It requires capturing actual traffic patterns from the production environment over a sufficient observation window — typically thirty to ninety days, long enough to include monthly batch processes, quarterly integrations, and any periodic communication that does not occur daily. This capture can be accomplished through network flow analysis tools, endpoint telemetry collection, or purpose-built segmentation planning platforms that automatically infer dependencies from observed traffic.

The output of the mapping phase is not a network diagram. It is a communication matrix: a structured record of which systems communicate with which other systems, using which protocols, on which ports, at what frequency. That matrix becomes the policy baseline. Every entry in it either survives into the new segmented environment as an authorized rule or is challenged: is this communication still needed, could it be redesigned to cross a less sensitive path, or is it a legacy artifact that should be eliminated? The challenge process itself is valuable — it frequently surfaces unnecessary connections that represent attack surface independent of any segmentation decision.

Diagram 5.1 – Communication Dependency Mapping Process

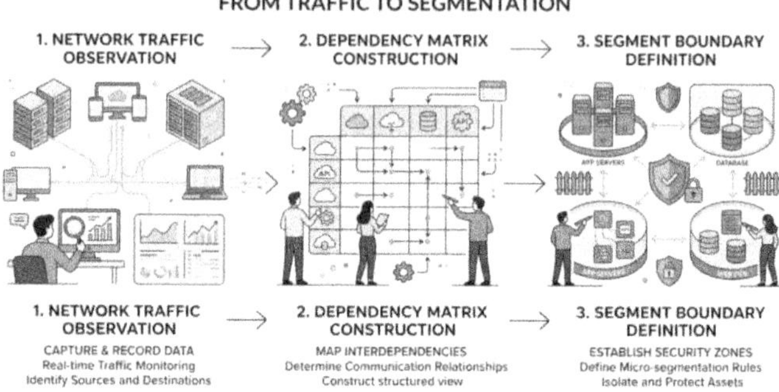

5.1.2 Identity-Based vs. Network-Based Segmentation Policies

Traditional network segmentation uses network constructs — VLANs, subnets, firewall rules based on IP addresses and port ranges — to define zones and control traffic between them. This approach has the advantage of familiarity and broad infrastructure support. It has the disadvantage of being tied to network topology: when a workload moves to a different host, its network-based policy may not follow it, because the policy is defined in terms of the address space it occupied, not the identity of the workload itself.

Identity-based segmentation policies attach access rules to the workload or user identity rather than to the network address. A containerized application carries its policy with it regardless of which host it runs on, which cloud region it is deployed in, or what IP address it is assigned. When two workloads need to communicate, the policy engine evaluates their identities and the communication request against the current policy, not the network topology. This model is inherently more portable, more consistent across hybrid and multi-cloud environments, and more resilient to the address changes that occur naturally in dynamic infrastructure.

The manager's evaluation criterion is operational sustainability. Network-based segmentation can be implemented faster because it works with existing infrastructure, but maintaining it accurately as the environment evolves requires sustained engineering effort. Identity-based segmentation requires a mature identity infrastructure as a prerequisite but produces policy that is inherently self-consistent as workloads move. For organizations early in their zero trust journey, a hybrid approach — network-based controls for initial segmentation with a migration pathway toward identity-based policy — often provides the most practical path forward.

5.1.3 Sequencing Implementation to Avoid Operational Disruption

Microsegmentation projects that attempt to enforce policy across the entire environment simultaneously almost always fail. They create a flood of policy violations — legitimate traffic that the new rules block — that operations teams cannot triage quickly enough to distinguish from genuine attacks. The result is either a rollback of the entire policy or a cascade of exceptions that hollows out the segmentation before it is ever fully in place.

The sequencing principle for microsegmentation is to begin in monitoring mode rather than enforcement mode. All policy rules are loaded into the environment, but violations generate alerts rather than blocking traffic. Operations teams use the monitoring period to validate that the policy allows all legitimate traffic and to identify rules that need adjustment. Only after the policy has been stable in monitoring mode — typically for two to four weeks per segment — does it move to enforcement mode. The transition is segment by segment, not environment-wide.

Priority sequencing should prioritize the highest-value assets. The systems that contain the data the organization would be most harmed by losing — customer records, intellectual property, regulated data, critical operational systems — should be the first to receive enforced segment boundaries, even if they are not the easiest to segment. The segmentation project that begins with the least sensitive systems and works toward the most sensitive ones creates protection where it is least needed and delays it where it is most urgent.

5.2 Software-Defined Perimeters: Abstracting Access from Physical Network Topology

Software-defined perimeters take the principle of network segmentation to its logical conclusion: rather than placing controls on the paths through the network, they make the resources themselves invisible until a requesting identity has been verified. An unauthenticated user scanning the network for accessible systems finds nothing — not a login page, not a port response, not a DNS entry. The resource does not exist from the network's perspective until the SDP controller has authenticated and authorized the requester and established a session specifically for that interaction.

The model derives from the concept of a dark network — one in which resources do not advertise their presence to unauthorized observers. The attack surface shrinks dramatically when the first step of any attack — reconnaissance, scanning, discovery — returns no results. A dark network cannot be mapped by an attacker who has not first authenticated, and authentication in an SDP model happens before any network access is established.

Diagram 5.2 – Software-Defined Perimeter Architecture

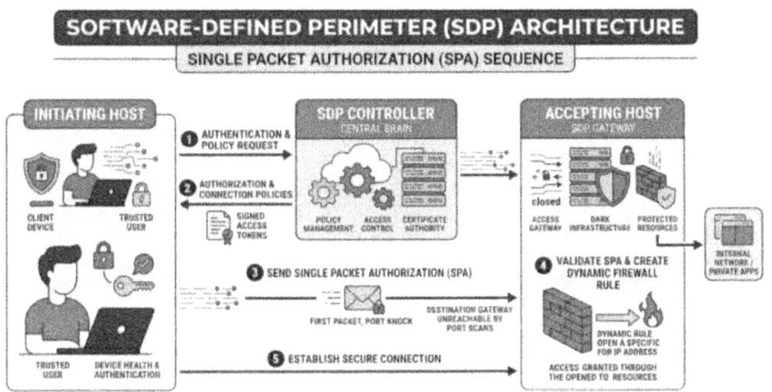

5.2.1 How SDP Controllers Gate Connectivity at the Session Level

The SDP controller is the decision point through which all access requests must pass. When a user or workload requests access to a resource protected by the SDP, the initiating client sends a cryptographically signed authorization request to the controller before attempting any network connection to the resource. The controller evaluates the request against identity, device posture, and policy — the same signals that any zero trust access decision requires — and, if approved, sends an authorization to both the requesting client and the accepting host to establish a direct, encrypted session between them.

The critical architectural property of this model is that the controller orchestrates the session but does not carry the traffic. Once the session is established, data flows directly between the client and the resource through an encrypted tunnel. The controller is not in the data path and therefore is not a bottleneck or a single point of traffic failure. It is a single point of authorization policy, which is both its strength and its architectural risk: the controller must be highly available, because its unavailability means no authorized sessions can be established.

For the manager, the SDP model has an important operational implication: users and workloads do not connect to the network and then access resources. They request specific resources, receive authorization for those specific resources, and establish sessions only to what they are authorized to reach. This is fundamentally different from VPN-based remote access, in which the user connects to the network and can then attempt to reach any system the VPN policy permits — which, in many organizations, is far more than the user legitimately needs.

5.2.2 Dark Network Principles and Pre-Authentication Invisibility

The dark network principle requires that resource endpoints not respond to any network traffic — ping, port scan, TCP connection attempt — from an unauthenticated source. This is implemented through a mechanism called Single Packet Authorization, or SPA, in which the client sends a single cryptographically signed UDP packet to the accepting host before any connection is attempted. The accepting host, which otherwise drops all unauthenticated traffic at the kernel level, examines the SPA packet, validates its signature and authorization status, and only then opens the port to the specific requesting address for a limited time window.

From the perspective of a network scanner, the SPA mechanism is invisible. The host appears to have no open ports. No service banners are returned. No login prompts are presented. The attack surface available to a pre-authentication attacker is, in practical terms, zero. This is not a firewall rule that blocks certain traffic — it is an architectural property in which the network infrastructure itself does not acknowledge the existence of the resource until authorization is confirmed.

The organizational implication is significant: a dark network cannot be breached through network reconnaissance because network reconnaissance produces no results. The attacker who has obtained credentials but has not reached the SDP authorization layer still cannot reach the resource. The credentials alone are insufficient — the device must also be authorized, the session must comply with policy, and the authorization must be granted before any network path exists to exploit.

5.2.3 SDP in Multi-Cloud and Hybrid Environments

The appeal of SDP architecture is particularly strong in multi-cloud and hybrid environments because it decouples access control from the network topology of any specific cloud provider. An organization with resources in three cloud environments and an on-premises data center does not need to maintain a consistent network segmentation policy across four different networking models. The SDP controller enforces a unified access policy regardless of where the resource lives, because the authorization decision happens at the identity and device level, not at the network level.

The implementation challenge in multi-cloud environments is controller placement and federation. The controller must be reachable by all clients regardless of where they are connecting from, must have access to identity and device posture signals from across the environment, and must be able to authorize sessions to resources in each cloud environment without requiring traffic to traverse a single choke point. Distributed controller deployments with synchronized policy state address this requirement, but they introduce the operational complexity of keeping multiple controller instances consistent and available simultaneously.

For organizations that are actively expanding their cloud footprint, evaluating SDP architecture before completing the expansion is substantially easier than retrofitting it afterward. The time to establish the authorization model is when new environments are being provisioned, not after years of network-based access patterns have been established and dependencies have accumulated.

5.3 Governing East-West Traffic: The Battle Inside the Network

East-west traffic — the lateral communication between servers, services, and workloads within the internal network — is where post-breach propagation happens. It is also where traditional security architecture has the least visibility and the weakest controls. The traffic that flows north-south, from user devices to application servers, has historically been inspected at gateway devices. The traffic that flows between servers in the same data center, between microservices in the same Kubernetes cluster, between database replicas in the same availability zone — that traffic has historically moved unexamined.

A zero trust architecture treats east-west traffic with the same level of scrutiny as north-south traffic. Every workload-to-workload communication is authenticated, authorized, and logged. The mechanisms through which this is accomplished vary by environment, but the requirement is consistent: implicit trust between internal systems is eliminated, and every communication is a verified transaction.

5.3.1 Service Mesh Architectures and Mutual TLS for Workload-to-Workload Trust

In cloud-native environments built on containerized microservices, the service mesh has emerged as the primary mechanism for enforcing east-west traffic policy. A service mesh inserts a proxy sidecar alongside each service instance. All traffic to and from the service instance passes through the sidecar proxy, which handles authentication, authorization, encryption, and telemetry collection. The service itself communicates as if it is on an open network; the mesh handles all the security properties transparently.

Mutual TLS is the authentication mechanism that service meshes typically use for workload-to-workload trust. In a mutual TLS session, both the client and server present certificates to authenticate their identities. Neither service accepts traffic from a counterpart that cannot present a valid certificate. A service mesh certificate authority issues the certificates, rotated automatically at defined intervals, and scoped to specific service identities rather than to network addresses. A compromised service cannot impersonate another service because it holds only the certificate for its own identity.

The manager's evaluation question for service mesh architecture is coverage: which services are inside the mesh, and which are not? A partial service mesh deployment that covers 80% of services in a cluster leaves 20% communicating without the enforcement properties the mesh provides. Attackers who understand the environment will route their lateral movement through the unprotected services. Coverage completeness is as important as the technical quality of the mesh implementation.

Diagram 5.3 – Service Mesh East-West Traffic Control

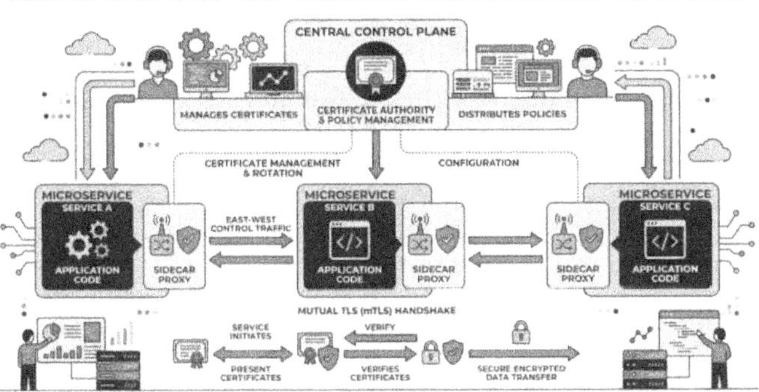

5.3.2 Detecting Lateral Movement Through Traffic Behavioral Analysis

Policy enforcement limits what lateral movement can accomplish. Behavioral analysis provides the means to detect it when it occurs despite policy controls. The two disciplines are complementary: policy is the barrier, detection is the signal that the barrier has been reached or bypassed. Neither alone is sufficient. A policy without detection leaves the organization blind to breach attempts. Detection without policy gives the attacker freedom to operate while the organization watches.

Traffic behavioral analysis in a segmented network leverages the observation points that segment boundaries create. Every boundary crossing is a policy decision that generates a log entry: allowed or denied, source identity, destination identity, protocol, volume, time. Over time, these log entries establish a behavioral baseline for the environment. The monitoring tools that analyze this baseline can surface anomalies: a service that normally communicates with three internal endpoints suddenly contacting fifteen; a workload that has never generated traffic toward the domain controller initiating queries against it; a batch process running eight hours ahead of its normal schedule and generating three times the typical data volume.

The challenge of behavioral analysis is alert calibration. An environment that generates thousands of anomaly alerts per day produces a team that learns to ignore them. The behavioral baselines must be sufficiently precise, and the anomaly thresholds sufficiently well-tuned, to produce a signal-to-noise ratio that security operations staff can act on. This calibration work is ongoing — the baseline must evolve as the environment evolves — and it requires sustained investment in the analytics platform and the analyst capacity to review and refine the outputs.

Diagram 5.4 – Lateral Movement Detection Through Behavioral Baselining

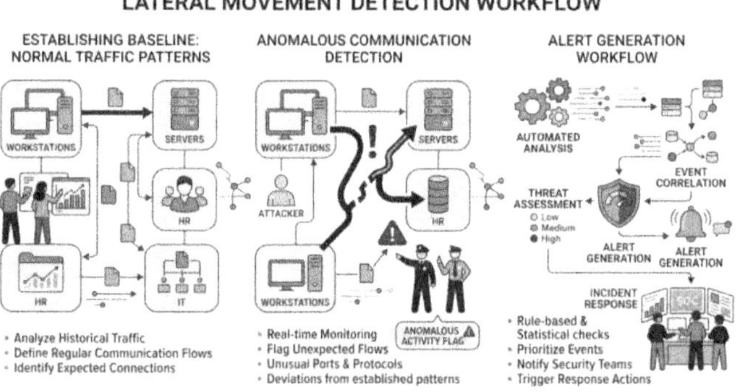

5.4 Segmentation in Non-Standard Environments

The microsegmentation frameworks described in the preceding sections were designed primarily for IT environments: server workloads, cloud services, user endpoints, and application tiers. They assume that components of the environment run modern operating systems, can install software agents, and can participate in certificate-based authentication. Many enterprise environments include components that meet none of those assumptions. Operational technology environments, cloud-native container platforms, and multi-cloud deployments each present segmentation challenges that require distinct approaches.

5.4.1 Operational Technology and Industrial Control System Considerations

Operational technology environments — the systems that control physical processes in manufacturing plants, power utilities, water treatment facilities, and similar industries — present a segmentation challenge that is qualitatively different from IT environments. OT systems often run proprietary operating systems that cannot be patched, modified, or have software agents installed without voiding vendor support or disrupting critical physical processes. Many were designed with decades-long operational lifespans and with network isolation as their primary security assumption — the Purdue model of OT security presumed that the systems would be air-gapped from IT networks and from the internet.

That assumption has not survived the convergence of IT and OT networks driven by operational efficiency requirements and remote management capabilities. OT systems that were once isolated now share network infrastructure with enterprise IT, creating lateral movement pathways from a compromised IT workstation to a programmable logic controller that operates physical equipment. The security consequences of that movement are not a data breach — they are the disruption or manipulation of physical processes with potential safety implications.

Zero trust segmentation in OT environments operates primarily through network controls at the OT-IT boundary, rather than through endpoint controls on OT devices themselves. The Purdue model's zone structure remains relevant, but the permeability of zone boundaries must be managed through policy-based enforcement points rather than through the assumption of physical isolation. Unidirectional security gateways — devices that allow data to flow in one direction only — provide a mechanism for OT systems to transmit telemetry to IT monitoring platforms without creating a bidirectional network path that could be exploited. The manager overseeing an environment that includes OT systems should ensure that the IT/OT boundary is explicitly mapped, actively monitored, and subject to the same policy rigor as any other segment boundary.

Diagram 5.5 – IT/OT Segmentation Boundary Architecture

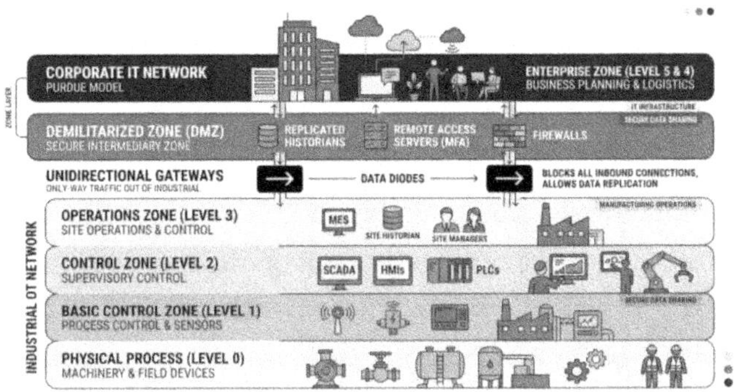

5.4.2 Cloud-Native Workloads and Container Network Policies

Container orchestration platforms introduce a segmentation environment that is highly dynamic — workloads start, stop, scale, and move continuously — and where the network addressing model is entirely virtual. Containers in a Kubernetes cluster communicate through an internal overlay network where addresses are assigned dynamically and can change whenever a workload is rescheduled. A firewall rule based on IP addresses in this environment is valid for the lifespan of a pod instance, which may be minutes.

Network policies in Kubernetes provide the mechanism for implementing microsegmentation at the container level. A network policy is a specification that defines which pods can communicate with which other pods, based on pod labels, namespace labels, and port specifications — not IP addresses. When a network policy is applied to a namespace or a set of pods, the container networking layer enforces it: traffic that does not match an allow rule is dropped by default. The policy travels with the workload identity, not with its address.

The default state in most Kubernetes deployments is the opposite of this. All pods in a cluster can communicate freely with all other pods unless a network policy explicitly restricts them. Enabling network policies requires both the policy specifications and a network plugin that enforces them. Organizations that have deployed Kubernetes without network policies have flat container networks — equivalent to the flat enterprise networks that microsegmentation is intended to replace, but with even less visibility because the traffic is entirely virtual and does not traverse physical network devices where it could be captured.

5.4.3 Managing Segmentation Policy Across Multiple Cloud Providers

Each major cloud provider has its own native security group model, its own virtual private cloud architecture, and its own set of controls for managing traffic between workloads. These models are not interoperable. A security group rule in AWS does not translate directly to a network security group in Azure. An organization that operates workloads in both environments cannot manage a single policy and apply it uniformly — it must maintain parallel policy definitions that express the same intent in platform-specific syntax.

Cloud-native policy management at multi-cloud scale requires either a cloud security posture management platform that can translate and apply policies across providers, or a network overlay architecture that abstracts the cloud providers' native networking and enforces a unified policy at the overlay level. Neither approach eliminates complexity — they relocate it. CSPM platforms add a management layer that must be maintained and kept up to date with each provider's API changes. Network overlays add latency and operational overhead to every workload communication.

The manager's governance requirement for multi-cloud segmentation is a policy audit capability that can answer, for any given workload in any cloud environment: what other resources can this workload reach, and what other resources can reach it? Without that audit capability, the segmentation policy exists on paper but cannot be verified in practice. Cloud configurations drift, provider defaults change, and manual exceptions accumulate. The audit capability is the feedback mechanism that keeps the policy honest.

Diagram 5.6 – Multi-Cloud Segmentation Policy Management

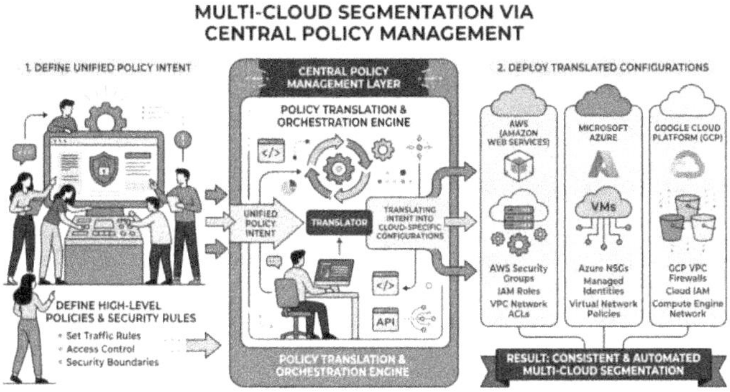

5.5 Manager's Checklist: Microsegmentation and East-West Control

Use this checklist to evaluate the operational maturity of your network segmentation program within a zero trust framework.

Confirm that a communication dependency mapping exercise has been completed for the highest-sensitivity environments, capturing actual traffic patterns over at least thirty days before any segmentation policy is enforced.

Verify that your segmentation implementation plan sequences enforcement by asset sensitivity — protecting the most critical systems first — rather than by implementation convenience.

Assess whether your current segmentation approach is network-based, identity-based, or hybrid, and confirm that a migration pathway exists toward identity-based policy as the environment matures.

Confirm that east-west traffic between server workloads is logged at segment boundaries and that a baseline behavioral profile exists against which anomalies are actively detected.

For cloud-native environments, verify that Kubernetes network policies are enabled and enforced in all namespaces that host production workloads, and that a policy audit capability confirms actual enforcement.

For OT/ICS environments, confirm that the IT/OT boundary is explicitly mapped, that all boundary crossing points are policy-enforced rather than assumed-isolated, and that OT traffic is included in security operations monitoring.

Assess multi-cloud segmentation coverage by confirming that a policy audit mechanism can enumerate reachability for any workload across all cloud environments in use.

Review the monitoring mode to enforcement mode transition record for each segment: confirm that no segment moved to enforcement without completing a validation period, and that exception rates are tracked and reviewed.

5.6 What This Means for Your Program

Network microsegmentation is the architectural response to the question that every post-breach investigation asks: why could the attacker reach that system from where they started? The answer to that question, in a flat network, is always the same: because nothing prevented them. Segmentation changes the answer by ensuring that reaching any system requires traversing a policy decision that either permits or denies the connection based on explicitly verified identity and authorization.

The manager's challenge in this domain is not technical — the tools for segmentation are mature and widely available. The challenge is organizational. Dependency mapping requires access to production traffic data that operations teams may be reluctant to share. Segment boundaries introduce operational changes that application teams must test and validate. The transition from monitoring to enforcement mode requires coordination across multiple teams and a shared commitment to accepting the friction of the change. These are coordination problems, not engineering problems, and they require manager-level sustained attention.

The program that treats microsegmentation as a one-time project will find that its policies drift out of alignment with the environment within months. New systems are provisioned, new integrations are established, and exception handling accumulates without a discipline for returning exceptions to the base policy. Segmentation is sustained by an operational cadence: regular policy reviews, dependency mapping refreshes as the environment changes, and a clear owner for the enforcement state of each segment. That cadence is the manager's responsibility to establish and protect.

6 Protecting the Asset That Actually Matters: Data Classification, Encryption, and Access Governance

A federal contractor discovered, during a routine storage audit, that an AI-enabled productivity tool had been indexing internal documents to build personalized response models for employees. The documents it had indexed included controlled unclassified information subject to handling restrictions, personally identifiable information subject to privacy agreements, and internal communications about ongoing contract negotiations. It had authorized the tool as a productivity application. No one had evaluated it as a data access event. No classification label on the files had triggered a policy response, because no consistent classification system existed. The data was accessible, the tool was authorized, and the exposure was complete before anyone thought to ask the question: what data is this application reaching, and should it have access to it?

Security architectures spend significant effort on who is accessing systems and how. They spend comparatively little effort on what data those systems contain and what the access decision actually means for that data. Zero trust repositions data at the center of the model. The access policy is not about users reaching servers — it is about principals reaching data under specific conditions. Until an organization knows what data it has, where it lives, and what its sensitivity requires in terms of access conditions, it cannot write a policy that means anything.

The manager's case for investing in data governance infrastructure is not compliance. Compliance frameworks establish data handling requirements, but they define a floor — the minimum actions required to avoid regulatory exposure. The zero trust case for data governance is operational: access policy without classification data is structurally incomplete. When the policy engine evaluates an access request, it needs to know not just who is requesting access and from what device, but what the data's sensitivity level requires in terms of authentication strength, session controls, and logging fidelity. Without that classification signal, the policy engine is making decisions without one of its primary inputs.

The workflow-level impact appears most clearly in regulated industries. A healthcare organization that cannot identify which records contain protected health information cannot enforce the access controls that HIPAA requires at the data level — it can only enforce them at the application level, which is a weaker and less precise control. A federal agency that cannot tag controlled unclassified information cannot enforce the handling rules that govern its transmission and storage. Classification is not the compliance team's administrative function. It is the architecture team's primary input for designing access policies.

The commitment this chapter is asking managers to make is to treat data classification as an architectural discipline, not an annual inventory project. Classification must be current, accurate, and machine-readable — tagged in a form that the access control infrastructure can act on, not just recorded in a spreadsheet that humans consult during audits. That discipline requires investment in tooling, processes, and ownership distinct from anything most organizations have historically applied to data governance.

6.1 Classification as Architecture Input, Not Compliance Checkbox

Data classification systems designed for compliance have a characteristic failure mode: they create categories that map to regulatory requirements rather than to access control decisions. The resulting taxonomy may satisfy an auditor's checklist while providing no actionable signal to the access policy infrastructure. A classification that marks a file as "confidential" tells the auditor that the organization acknowledges the file's sensitivity. It does not tell the access control system who should reach it, under what authentication conditions, from what device states, or what logging the access event should generate.

Effective classification for zero trust requires a taxonomy designed from the access policy backward. Each classification level must map directly to a set of policy conditions: minimum authentication strength required, device posture requirements, network location restrictions, session recording requirements, and logging fidelity level. When an access request arrives for data tagged at a given classification level, the policy engine can immediately determine what it needs to verify before granting access.

6.1.1 Building a Classification Taxonomy That Access Policy Can Actually Use

A workable zero trust classification taxonomy requires four to five tiers, not because regulatory frameworks demand that number, but because access policy needs that level of granularity to express meaningfully different control requirements. A two-tier system — public and restricted — cannot express the difference between data that requires strong authentication and data that requires phishing-resistant authentication plus session recording plus restricted network origin. A ten-tier system creates categorization complexity that exceeds any team's ability to apply consistently.

Each tier in the taxonomy should be defined by two things: the nature of the data it covers, and the policy conditions that access to it triggers. The definition must be specific enough that a content owner can apply it to a document without ambiguity, and specific enough that the access policy engine can translate it into enforcement rules without interpretation. Vague tier definitions — "sensitive information of a proprietary nature" — generate inconsistent classification and unenforceable policy. Precise tier definitions — "data whose unauthorized disclosure would enable competitive harm or regulatory penalty" — give classifiers and policy systems a shared reference point.

The taxonomy design process should include representatives from the access control team, the compliance function, and the business units that own the most sensitive data categories. The compliance team can identify the regulatory requirements the taxonomy must accommodate. The access control team can specify what policy conditions are available for each tier to trigger. The business unit representatives can identify the data categories that the taxonomy must cover and validate that the tier definitions are unambiguous when applied to actual content. A taxonomy designed by one of these groups in isolation will fail to serve the others.

Diagram 6.1 – Data Classification Taxonomy Mapped to Policy Controls

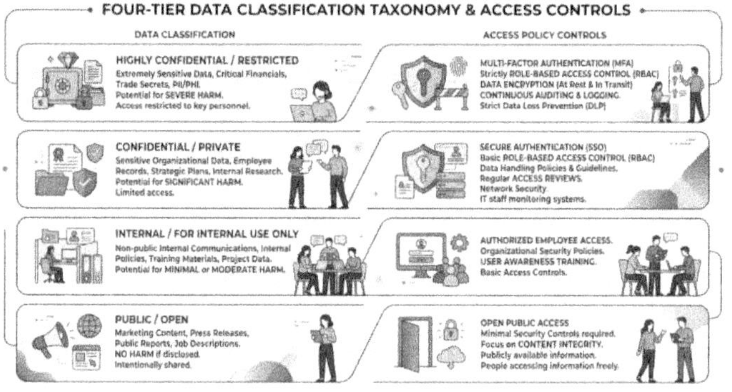

6.1.2 Automated Discovery and Tagging at Scale

Manual classification does not scale in any meaningful environment. An organization with hundreds of terabytes of documents, emails, and database records distributed across cloud storage, on-premises file shares, collaboration platforms, and archive systems cannot rely on employees to tag every item accurately. Human classifiers are inconsistent, biased toward under-classification when classification imposes access friction, and absent for the large volumes of data that are never touched after creation.

Automated data discovery and classification tools address this gap by scanning content at rest — in storage repositories, cloud services, databases, and collaboration platforms — and applying classification labels based on content analysis. Pattern matching identifies structured sensitive data, such as social security numbers, credit card numbers, health record identifiers, and similar formats that appear consistently and can be detected with high confidence. Machine learning models extend this capability to unstructured content: contracts, clinical notes, internal communications, and other documents where sensitivity is expressed through context rather than format.

Automated classification is not a set-and-forget process. The initial scan establishes a baseline, but the classification infrastructure must continue scanning as new data is created and as existing data is modified. A document that was classified as internal upon creation may become regulated data when it is updated to incorporate patient information. A file that was marked as publicly shareable may be reclassified when it is moved to a folder that contains sensitive content. The classification system must detect these changes and update labels accordingly, because an access policy that enforces controls based on stale classification data is not enforcing the correct policy.

6.1.3 Maintaining Classification Accuracy as Data Moves and Ages

Data does not stay where it was classified. It is copied to collaboration platforms, emailed to external parties, archived to cold storage, migrated to new cloud environments, and duplicated for backup purposes. Each movement is an opportunity for the classification label to be lost, overwritten, or become disconnected from the access controls associated with it. A document classified as highly sensitive in the enterprise content management system that is emailed to a collaboration platform may arrive there without its classification metadata, effectively appearing as unclassified content to the receiving system's policy engine.

Persistent classification labels — metadata embedded in the document itself rather than stored only in the repository — address the portability problem. When a document carries its classification label as intrinsic metadata, that label travels with the file regardless of where it is moved. Email and collaboration platforms that integrate with the enterprise classification system can read the embedded label and apply the appropriate access controls in the destination environment. The label serves as a trust signal that the receiving system can act on without re-examining the content.

Data aging presents a related challenge: data sensitivity changes over time, and classification labels that were accurate at creation may be incorrect years later. A contract negotiation file that was classified as highly sensitive during the negotiation period may become a historical record once the contract is executed, with different access requirements. An employee complaint file that was restricted during an active HR investigation may be subject to different access rules once the investigation is closed and the record is retained for legal purposes. Classification governance must include a review cadence for aged data — not a full manual review of every file, but a policy-driven reassessment process that identifies data whose classification should be reviewed based on age, status change, or regulatory retention schedule.

Diagram 6.2 – Classification Label Persistence Across Data Movement

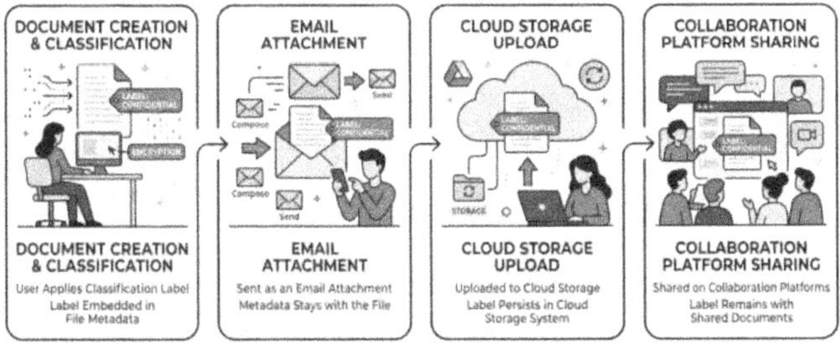

6.2 Encryption That Works Across the Full Data Lifecycle

Encryption is the last line of defense for data that has left its authorized access context. When an attacker exfiltrates a file, when a storage device is physically removed from a data center, when backup media is lost in transit, when a cloud misconfiguration exposes a storage bucket to the public internet, encryption is the control that determines whether the exposed data is readable. It is not a substitute for access control, but it is the control that remains effective when access control has failed.

The manager's framing for encryption decisions is risk coverage across three distinct states: data at rest (stored on disk, in cloud storage, in backup media), data in transit (moving across networks between systems), and data in use (actively being processed by an application). Each state presents different encryption challenges and requires different technical approaches. An organization that encrypts data at rest and in transit but has no controls on data in use leaves a significant exposure window during the processing phase.

6.2.1 Key Management as the Critical Dependency

Encryption is only as strong as the key management that supports it. Data encrypted with a well-designed algorithm is computationally infeasible to decrypt without the key. Data encrypted with a well-designed algorithm whose keys are stored in plaintext alongside the encrypted data is effectively unencrypted. Key management is the discipline of generating, storing, distributing, rotating, and retiring encryption keys in a way that maintains the separation between the encrypted data and the capability to decrypt it.

Hardware security modules provide the highest assurance for key storage and cryptographic operations. An HSM is a tamper-resistant device that generates and stores keys in protected hardware, performs cryptographic operations internally without exposing key material to software layers, and provides an auditable record of every operation. Keys stored in an HSM cannot be extracted in plaintext — the key material never leaves the device. This property is critical for the highest sensitivity data categories, where the risk of key compromise is itself a material security event.

Cloud key management services provide a more operationally accessible path for organizations without dedicated HSM infrastructure. Each major cloud provider offers a managed key management service that stores keys in hardware-protected enclaves and integrates with the provider's storage, database, and application services. The operational advantage is significant: automated key rotation, centralized key lifecycle management, and integration with cloud-native access controls. The strategic consideration is cloud provider dependency: keys managed by a cloud provider create a dependency on that provider for decryption of all data encrypted under those keys. Organizations managing data subject to regulatory requirements, or data that could be sensitive in the context of a cloud provider relationship, should evaluate whether customer-managed keys with bring-your-own-key capabilities are appropriate.

6.2.2 Encryption at Rest: Storage, Backup, and Archive Strategies

Encryption at rest protects data from physical access to storage media — a hard drive removed from a decommissioned server, a backup tape lost in transit, a laptop stolen from a car. It also protects against unauthorized access to storage systems where the attacker has bypassed access controls to reach the storage layer directly rather than through the application. The data on the disk is unintelligible without the key, which the attacker does not hold.

Transparent database encryption — in which the database engine handles encryption and decryption automatically, without requiring application changes — is the most operationally practical approach for production databases. The data is encrypted on disk and decrypted in memory when queried through the authorized application. The limitation of transparent encryption is that it protects against physical storage access but does not protect against an attacker who has compromised the application layer: the application holds the capability to decrypt, so an application-level compromise is equivalent to a plaintext compromise.

Backup and archive encryption is an area where many organizations have significant coverage gaps. Production data may be encrypted at rest, but backup copies are sometimes stored in formats designed for rapid restoration rather than security. Archive storage, particularly older archives that predate the organization's encryption policy, may contain years of unencrypted data that represents the most complete picture of the organization's historical records. A complete encryption program must include an assessment of backup and archive coverage, along with a remediation plan for any unencrypted repositories containing data subject to classification-based handling requirements.

Diagram 6.3 – Encryption Coverage Across Data States and Locations

6.2.3 Encryption in Transit and the Inspection Dilemma

Transport Layer Security has become the default standard for encrypting data in transit across both internal and external network paths. TLS ensures that traffic between a client and a server — or between two services — cannot be read or modified by an intermediary without detection. Its near-universal adoption has eliminated the plaintext exposure that made network sniffing a primary attack technique in earlier eras.

The inspection dilemma arises from an architectural tension between encryption and visibility. Security operations teams rely on the ability to inspect network traffic for indicators of malicious behavior, such as data exfiltration patterns, command-and-control communications, and anomalous protocol usage. When that traffic is encrypted end-to-end, it is invisible to network-based inspection tools. The same encryption that protects legitimate traffic from interception also protects malicious traffic from detection.

TLS inspection — a technique in which the inspection device terminates the encrypted connection, examines the plaintext, and re-encrypts it before forwarding it — is the conventional response to this dilemma. It restores visibility to encrypted traffic but requires the inspection device to hold a certificate trusted by both endpoints, creating a man-in-the-middle position that, if the inspection device is compromised, exposes all traffic it processes. For organizations subject to regulatory requirements that restrict who may access certain data, TLS inspection of connections to systems that handle that data may also create compliance exposure. The inspection dilemma requires a deliberate policy decision — not a default — about which traffic categories are subject to inspection, what safeguards protect the inspection infrastructure, and how compliance implications are addressed.

6.3 Access Governance: Keeping Permissions Honest Over Time

Access permissions drift. A user who was granted access to a sensitive data repository for a six-month project retains that access after the project ends because no one filed a ticket to remove it. A role defined for a specific business function accumulates permissions as new requests are approved, without evaluating the full picture of what the role already has. A service account created for a narrow integration purpose is reused for a second integration that requires broader access, and the original narrow permissions are expanded rather than creating a second, appropriately scoped account. Over time, the gap between what each identity should be able to access and what it actually can access grows consistently in one direction: more access than required.

Access governance is the discipline of closing that gap through structured, recurring processes that compare current permissions against current legitimate need and eliminate the difference. It is the operational mechanism that maintains least privilege over time rather than just being enforced at provisioning. Without it, even an organization that provisions access precisely will find its access control posture degrading continuously as the environment changes.

6.3.1 Entitlement Reviews That Surface Real Risk Rather Than Rubber Stamps

Access review programs are in place in most regulated organizations. In most cases, they function as certification exercises: managers receive lists of their employees' access, click "Approve" on each item, and submit the completed form. The review generates a compliance artifact demonstrating that access was reviewed. Still, it provides no meaningful assurance that access is appropriate, because the manager who receives the list has no context to evaluate whether each access grant is still justified, no consequence for approving inappropriate access, and no time budgeted actually to investigate items that seem inconsistent.

Effective entitlement reviews require a different design. The review interface should present each access grant with the business context that justifies it — the original approval reason, the last time the access was used, the data sensitivity level of the resource — so that the reviewer can make an informed judgment rather than a reflexive confirmation. Unused access should be flagged prominently: an access grant that has not been exercised in ninety days requires positive justification for continuation, not passive approval. High-risk access combinations — rights to modify access permissions combined with rights to sensitive data — should be routed to a secondary reviewer, because a manager whose primary obligation is operational continuity may not recognize the risk profile of a specific access combination.

The cadence of reviews must be calibrated to the sensitivity of the access being reviewed. Privileged access reviews should occur quarterly or continuously through automated monitoring. Access to highly classified or regulated data should be reviewed at least semi-annually. Standard access reviews can be annual for stable roles with well-defined access requirements. A flat review cadence, applied uniformly across all access categories, either creates unsustainable review volume for low-risk access or inadequate review frequency for high-risk access.

6.3.2 Attribute-Based Access Control and Dynamic Permission Assignment

Role-based access control assigns permissions to roles and assigns roles to users. It is the dominant access control model in enterprise environments because it scales to large organizations. Rather than managing permissions for each user individually, the organization manages a set of roles and their memberships. The limitation is that roles are static: a user either has the role or does not, and the associated permissions apply in all contexts.

Attribute-based access control replaces the binary role assignment with a policy that evaluates multiple attributes at the time of each access request. The data's classification level, the user's current location, the device's posture state, the time of day, the purpose of the request, and the current risk signal from behavioral analytics — all of these attributes can be incorporated into the access decision, which is made dynamically at request time rather than assigned statically at provisioning time. The result is access that is precisely calibrated to the context of the specific interaction rather than to the general category of the user's role.

The operational challenge of ABAC is the complexity of policy management. A policy that evaluates six attributes across four classification levels for hundreds of resource types generates a policy set that is difficult to validate, audit, and troubleshoot. Organizations that move toward ABAC should begin with the highest-sensitivity data categories — where the additional policy complexity is justified by the access-control precision it provides — and expand from there, rather than attempting to implement attribute-based decisions across all access simultaneously. The governance requirement is a policy management discipline that tracks the rationale for each attribute condition, ensures that policy changes are reviewed before deployment, and provides a simulation capability that allows operators to test the effect of a new policy before it is enforced in production.

Diagram 6.4 – Attribute-Based Access Decision Evaluation

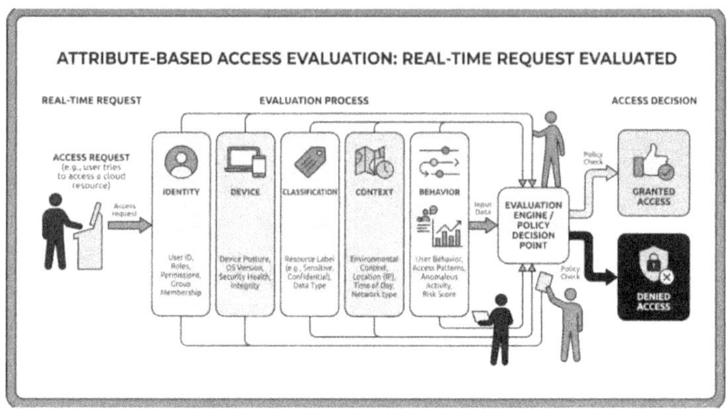

6.4 Data Protection in Regulated Environments

Regulated data environments layer sector-specific requirements onto the zero trust framework. The HIPAA requirements for protected health information, the federal government's handling standards for controlled unclassified information, and the data sovereignty requirements imposed by jurisdictions where cross-border data flows are restricted — each adds a dimension of specificity to the access governance model that the general zero trust architecture must accommodate. The challenge is satisfying these requirements without creating compliance-specific architectures that fragment the unified access control model.

The manager's orientation for regulated environments should be that compliance is a floor, not a ceiling. Rigorously applying zero trust principles will satisfy the technical requirements of most data protection regulations, because those regulations are concerned with the same outcomes: knowing where sensitive data lives, controlling who can access it and under what conditions, logging access events, and maintaining the integrity of the data. The additional burden of regulated environments is documentation and demonstrability — the organization must be able to show regulators that the controls exist, are correctly configured, and are actively enforced.

6.4.1 Healthcare: Satisfying HIPAA While Enabling Clinical Workflows

HIPAA's security rule establishes requirements for protecting electronic protected health information: access controls, audit controls, integrity controls, and transmission security. These requirements are consistent with zero trust principles, but applying them in a clinical environment requires careful calibration. Clinical workflows impose authentication demands that would be unworkable under standard enterprise authentication models. A physician moving between examination rooms may need to access patient records dozens of times per shift on shared workstations. An emergency department clinician may need to override access controls during a patient emergency more quickly than a standard step-up authentication flow allows.

The zero trust response to clinical workflow requirements is a context-aware access policy calibrated to the healthcare environment. Proximity-based authentication — using Bluetooth or RFID to authenticate a clinician to a nearby workstation based on their badge — provides fast, low-friction authentication appropriate for high-frequency access in clinical spaces. Break-glass procedures with mandatory logging and post-hoc review provide a compliant mechanism for emergency access that bypasses standard controls when patient care requires it. Role-based access, scoped to the treating clinician's active patient panel, restricts access to a relevant set of records rather than the entire patient database, thereby applying least privilege at the clinical workflow level.

Audit logging in healthcare environments serves both the access governance and regulatory demonstrability functions. Every access to a protected health information record should generate a log entry that includes the identity of the accessing user, the record accessed, the clinical justification, if provided, the time, the device, and the network location. These logs are the evidence the organization presents during a HIPAA audit, and they are the data that access governance reviews use to detect inappropriate access — the clinician who accessed a celebrity patient's record without a treating relationship, the administrative staff member who accessed records of former colleagues.

Diagram 6.5 – Zero Trust Access Controls for Clinical PHI Environments

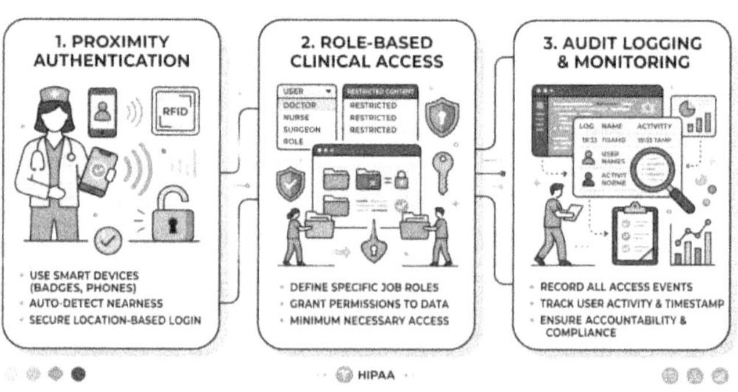

6.4.2 Federal Classified and Controlled Unclassified Information Handling

Federal agencies and contractors handling controlled unclassified information operate under a framework of specific handling requirements established by the National Archives and Records Administration's CUI program. CUI handling requirements specify how information must be labeled, stored, transmitted, and protected, and they impose compliance obligations on any organization that handles CUI under a federal contract. The NIST SP 800-171 standard translates these requirements into specific security controls for non-federal systems handling CUI.

The zero trust architecture for CUI environments must address three requirements that the general enterprise model may not handle by default: mandatory access control labels that restrict access based on clearance and need-to-know rather than role alone; audit log retention at durations specified by federal requirements, which may extend to three or more years; and boundary controls that ensure CUI does not flow to systems outside the authorized handling environment. Each of these requirements maps to specific capabilities in the zero-trust architecture — ABAC policies that incorporate clearance and need-to-know attributes, log retention policies enforced at the telemetry infrastructure level, and data loss prevention controls that intercept CUI exiting via email, cloud uploads, or portable media.

The operational challenge for contractors and agencies is maintaining separation between CUI-handling environments and general enterprise environments without creating two entirely parallel technology stacks. A unified identity platform that issues different credential types for different access contexts — standard enterprise credentials for routine business systems and additional assurance credentials for CUI access — reduces the operational overhead of maintaining separation while providing the control boundary that federal requirements demand.

6.4.3 Cross-Border Data Flows and Jurisdictional Complexity

Data sovereignty requirements — regulations that restrict the transfer of personal data across jurisdictional boundaries without specific legal safeguards — introduce a data-residency dimension that the zero-trust access governance model must track and enforce. The European Union's General Data Protection Regulation establishes the most widely applicable cross-border transfer framework: personal data of EU residents may only be transferred to jurisdictions that the European Commission has determined offer adequate protection, or under specific legal mechanisms, including standard contractual clauses, binding corporate rules, or approved certification frameworks.

For organizations with multinational operations, data residency compliance requires knowing where data is stored at any given moment and ensuring that access from outside the data's permissible jurisdiction is either technically prevented or conducted under an appropriate legal framework. In cloud environments, where data may be automatically replicated across geographic regions for performance and availability, the residency of any given piece of data can change without the data owner's explicit action. The access governance model must integrate residency tracking with the classification and access control infrastructure so that cross-border access requests to restricted data are evaluated against the applicable legal transfer mechanism before access is granted.

The manager overseeing data governance in a multinational environment should require a data residency register that identifies the storage location of each data category subject to cross-border restrictions, the legal transfer mechanism in place for each permissible cross-border flow, and the access control configuration that enforces the boundary for flows that are not permitted. That register must be actively maintained — cloud environments change, legal mechanisms expire, and regulatory interpretations evolve — and it must be accessible to the access policy infrastructure, not just to the compliance team.

Diagram 6.6 – Cross-Border Data Flow Governance Model

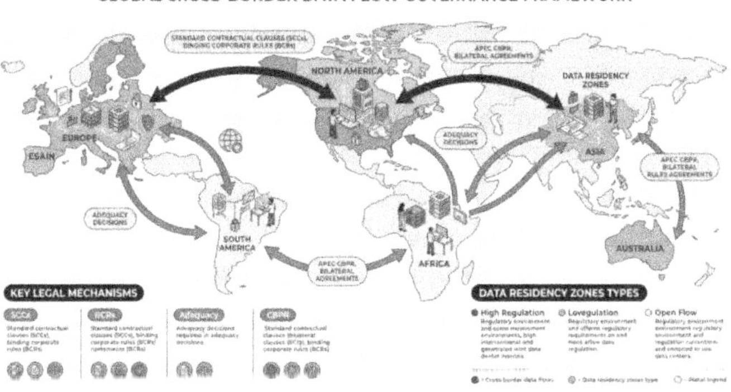

6.5 Manager's Checklist: Data Classification, Encryption, and Access Governance

Use this checklist to evaluate the operational readiness of your data protection program within a zero trust framework.

Confirm that a data classification taxonomy exists that maps directly to access policy conditions — authentication strength, device posture, logging requirements — for each tier, not just to regulatory categories.

Verify that the automated data discovery tooling is scanning all production data stores, including cloud storage, collaboration platforms, and databases, and that the scan results are used to drive classification label application rather than just to generate inventory reports.

Confirm that classification labels are embedded in documents as persistent metadata that travels with files when they are moved or shared, rather than stored only in the source repository.

Audit encryption coverage across data at rest, in transit, and in backup and archive storage — confirm that the coverage map identifies any repositories holding classified data that are not yet encrypted.

Verify that key management infrastructure — whether HSM-based or cloud-managed — provides hardware protection for keys governing the highest classification tiers, and that key rotation schedules are automated.

Review the most recent access certification cycle: confirm that unused access was flagged and required positive justification, that high-risk access combinations received secondary review, and that the exception rate and outcome were tracked.

For HIPAA-covered environments, confirm that break-glass access procedures are documented, logged, and subject to mandatory post-hoc review, and that audit logs are retained at the required duration with integrity controls.

For environments handling CUI or subject to cross-border data flow restrictions, confirm that a current residency register exists, that it is integrated with access policy enforcement, and that it is reviewed on a defined cadence by the compliance function.

6.6 What This Means for Your Program

Data protection is where zero trust architecture earns its name most directly. The architecture exists to protect assets — and data is the asset that attackers ultimately want. Identity controls determine who can initiate a request. Network controls determine whether that request can reach the system. Data classification, encryption, and access governance determine whether the data the attacker reaches when they breach those controls is readable, whether their access was logged, and whether the governance structure around that data was sufficiently mature to surface the anomaly before the damage was complete.

The manager who treats data classification as a compliance exercise and access governance as an annual checkbox will find that the identity and network controls they have invested in are protecting a core that has not been properly hardened. Access policy without classification input is structurally incomplete. Encryption without key management discipline provides false assurance. Access reviews without design intent behind them produce certification artifacts rather than assurance.

The practical starting point for most organizations is the classification inventory question: Does the organization know where its most sensitive data lives today, at this moment, across cloud environments, collaboration tools, backup systems, and legacy repositories? If the answer is uncertain, the data protection program begins there, not with an encryption platform selection or an access review tool implementation, but with the discipline of knowing what you have before deciding how to protect it. Every other investment in this chapter depends on that foundation.

7 From Whiteboard to Production: Phased Rollout, Maturity Models, and Measurable Progress

A federal agency CISO stood before her deputy secretary with a printout of the agency's Zero Trust roadmap — 18 months old, 22 action items, and a status column showing 11 items still marked "in planning." The technology was approved. The vendor was under contract. The budget had been allocated. What the roadmap lacked was a sequence, a set of measurable milestones, and a clear definition of what "done" meant for each phase. The program had stalled, not because Zero Trust was too complex, but because there was no structured way to navigate from the current state to the target state without losing the thread along the way.

This chapter provides that structure. It begins where most programs actually begin — not with a clean slate but with a technology estate that already has controls, dependencies, and constraints. The work is to map what exists, assess what it covers in Zero Trust terms, identify a feasible sequence of changes, and build measurement practices that let leadership track whether the transformation is actually reducing risk. Done correctly, this process turns a multi-year program into a series of navigable milestones, each of which delivers real security value and builds the organizational confidence needed to sustain the work.

Why it matters: Zero Trust deployments that lack a structured approach to sequencing and measurement tend to fail in one of two ways. The first is scope collapse — the program gets narrowed to a single technology layer, identity or network. It never integrates with the others, leaving the architecture incomplete and its gaps exploitable. The second is momentum loss — early phases take longer and cost more than projected, executive patience erodes, and the program is quietly deprioritized before it reaches the domains with the most exposure. Both failure modes are preventable when the implementation strategy is built on honest assessment, realistic sequencing, and metrics that demonstrate progress in terms that decision-makers actually understand.

The manager's role in this process is not to master every technical dependency — that is the engineering team's responsibility. It is to ensure that the program has a navigable structure, that each phase has defined entry and exit criteria, and that the reported metrics actually tell a security story rather than a technology deployment story. A dashboard showing that sixty percent of endpoints are enrolled in device management is less useful than one showing that policy enforcement is blocking unauthorized access attempts at a measurable rate. The distinction matters because it determines whether leadership can make informed decisions about pace, investment, and the acceptance of risk.

7.1 Diagnosing the Starting Point: Current-State Assessment and Gap Analysis

No Zero Trust program starts from zero. Every enterprise already has access controls, some form of identity management, network segmentation in at least a few places, and logging capabilities of varying quality. The task is not to set these aside and rebuild from scratch — it is to understand what they cover in Zero Trust terms, where they fall short, and what dependencies must be resolved before new capabilities can be layered on top of them.

Current-state assessment is therefore the first investment the program must make. It is also one of the most politically sensitive, because it surfaces uncomfortable truths about the gaps left by previous security investments. The goal is not to assign blame but to establish a shared factual baseline that the program can use to set priorities and sequence its work. When that baseline is built carefully and communicated clearly, it becomes the most useful tool the program has for driving budget decisions and justifying phase sequencing to skeptical stakeholders.

7.1.1 Mapping Existing Controls to Zero Trust Capability Areas

The first task is to map the organization's current controls against the five major Zero Trust capability areas: identity, devices, networks, applications and workloads, and data. For each capability area, the question is not whether any controls exist but whether those controls enforce policy in a way that is consistent with Zero Trust principles — explicit verification, least-privilege access, and breach assumption.

In practice, most organizations find that identity controls are partially mature, with strong authentication implemented for some user populations but not others, and machine identity management largely unaddressed. Network controls tend to reflect the old perimeter model — strong at the edge, flat internally. Application-level controls are often inconsistent across the portfolio, stronger for externally facing systems and weaker for internal tools. Data classification and access governance are typically the least mature capability areas. Device controls vary depending on whether the organization has invested in endpoint management, and usually show significant gaps for contractor-owned and unmanaged devices.

Mapping this landscape does not require a formal assessment tool, though several exist and can accelerate the process. It requires structured conversations with the teams responsible for each domain, a review of existing policy documentation, and enough technical sampling to validate what the policies say versus what is actually enforced in production. The output is a capability heat map — a clear picture of where the organization sits relative to Zero Trust requirements across all five pillars.

Diagram 7.1 – Zero Trust Capability Heat Map

Zero Trust Capability Heat Map

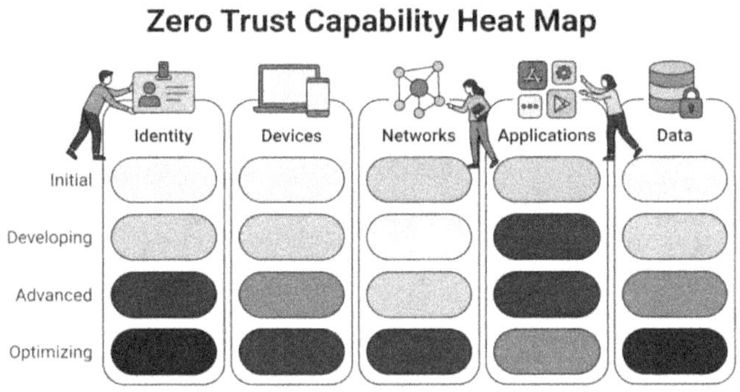

7.1.2 Identifying Quick Wins That Build Momentum Without Incurring Technical Debt

Not all capability gaps are equally complex to close. Some improvements can be made quickly, with existing tools and modest effort, and will produce meaningful security outcomes. Identifying these quick wins is strategically important because they demonstrate program value early, build confidence across stakeholder groups, and create operational familiarity with Zero Trust controls before the more complex phases begin.

Common quick wins include enforcing multi-factor authentication for all privileged accounts, implementing conditional access policies that block authentication from unregistered devices, revoking dormant accounts and service credentials that have not been used in the past ninety days, and segmenting the most sensitive application environments from the rest of the network using existing firewall capabilities. Each of these actions can typically be implemented within a single sprint cycle and produces measurable risk reduction within weeks.

The discipline required in this phase is to avoid accumulating technical debt in pursuit of speed. Quick wins that are implemented in ways that will need to be rearchitected later — such as conditional access policies that rely on IP address allowlists rather than device identity — can actually slow the program down by creating legacy configurations that must be unwound before later phases can proceed. Every quick win should be implemented in a way consistent with the target architecture, even if it does not yet fully realize that architecture.

7.1.3 Dependency Mapping: What Must Exist Before Each Next Step

Zero Trust capabilities are not independent — they have prerequisites. Device-based access policy requires a device management system that can produce the posture signals the policy engine needs. Microsegmentation requires a comprehensive map of application communication dependencies before segment boundaries can be drawn without breaking production workflows. Behavioral analytics requires a log-collection infrastructure with consistent signal formats for reliable anomaly detection.

Dependency mapping is the work of identifying these prerequisites and sequencing the program's phases accordingly. It is not glamorous work, but it is the single most common cause of implementation delays when skipped. Teams that proceed to policy configuration before identity federation is complete, or to network segmentation before application dependency mapping is finished, routinely encounter production incidents that force rollbacks and erode executive confidence in the program.

The output of dependency mapping is a sequenced phase plan in which each phase has explicit prerequisites — capabilities that must be in place and validated before the phase begins. This is distinct from a project timeline, which can be constructed top-down from a desired completion date. A dependency-driven sequence is built bottom-up from the environment's technical realities. It represents the minimum viable order in which work can be done without creating risk or rework.

Diagram 7.2 – Zero Trust Dependency Sequence Map

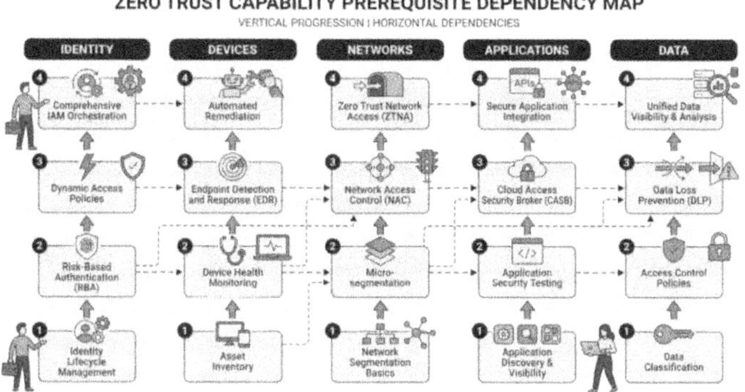

7.2 Maturity Models as Navigation Tools, Not Report Cards

Maturity models serve a specific function in a Zero Trust program: they provide a shared vocabulary for describing the current state and a common framework for discussing the target state. What they do not do — when used correctly — is tell an organization where it should aim. That decision is a function of risk appetite, regulatory obligations, and the specific threat environment the organization operates in, none of which are captured in a generic maturity scale.

The danger of treating maturity models as scorecards rather than navigation tools is that they create incentives to improve scores rather than reduce risk. An organization that achieves a "managed" rating across all five CISA pillars by implementing the minimum required controls for each level may have a strong assessment report and a significantly weaker security posture than an organization at an "initial" rating in two pillars that it has identified as its highest-risk domains and is investing in intensively.

7.2.1 CISA Zero Trust Maturity Model: Pillars and Progression Stages

The CISA Zero Trust Maturity Model organizes its assessment across five pillars — identity, devices, networks, applications and workloads, and data — and describes four progression stages for each: traditional, initial, advanced, and optimal. Each stage is characterized by increasingly automated, integrated, and dynamically enforced controls. At the traditional stage, controls are largely manual and perimeter-centric. At the optimal stage, policy decisions are driven by real-time signals, enforcement is continuous, and exceptions are logged and reviewed automatically.

The model is useful because it provides concrete, observable characteristics for each stage across each pillar. A team can read the description of "advanced" identity controls and, with relatively low ambiguity, determine whether their current identity program meets that description. This specificity makes the model a practical basis for gap analysis and phase planning, and a credible framework for communicating progress to federal oversight bodies and other external audiences familiar with the CISA model.

The model is less useful when it is used to set a single target stage across all pillars simultaneously. Different pillars may require different target stages depending on the organization's risk profile. A healthcare organization handling electronic health records may need to reach the optimal stage for data controls and an advanced stage for identity, while accepting an initial stage for operational technology network controls that cannot be segmented without clinical disruption. The model should be applied with this nuance, not as a uniform bar to clear.

7.2.2 Customizing a Maturity Target to Organizational Risk Appetite

Setting a maturity target is a risk management decision, not a technical one. It requires the security team to articulate the specific threats and failure scenarios that the organization is most exposed to, the regulatory requirements that establish minimum control floors, and the operational constraints — budget, staff, legacy infrastructure — that shape what is realistically achievable within a defined time horizon.

The output of this process is a target state profile: a maturity stage for each pillar that reflects the organization's specific risk calculus. This profile becomes the destination against which phase sequencing is planned, progress is measured, and exceptions are evaluated. It also provides the basis for a risk acceptance statement that documents which gaps the organization has chosen to accept and why — a document that is increasingly expected by regulators, auditors, and insurance underwriters.

Managers who participate in this process should bring the business context that security teams sometimes lack: which systems, data sets, and workflows represent the highest operational and reputational risk, what disruption the organization can absorb during transition phases, and where the organizational dependencies across business units create sequencing constraints that are not visible from the technology layer alone.

7.2.3 Using Maturity Assessments to Drive Budget Conversations

Maturity assessments are most valuable to managers when they are translated into budget terms. A gap between the current state and the target state in the identity pillar is not inherently persuasive to a budget committee. The same gap framed as "our current identity controls cannot distinguish a legitimate remote employee from an attacker using stolen credentials and closing that gap requires the following investments over the following timeline" is considerably more actionable.

This translation requires the security team to map each maturity gap to a specific risk scenario and a control capability that addresses it. The investment required to close the gap — in technology, staffing, and operational overhead — can then be compared to the risk reduction it produces and the cost of a breach that the current gap enables. This is the language of budget conversations, and it is the language in which Zero Trust investments must be articulated to compete successfully with other organizational priorities.

Diagram 7.3 – Maturity Target Profile by Pillar

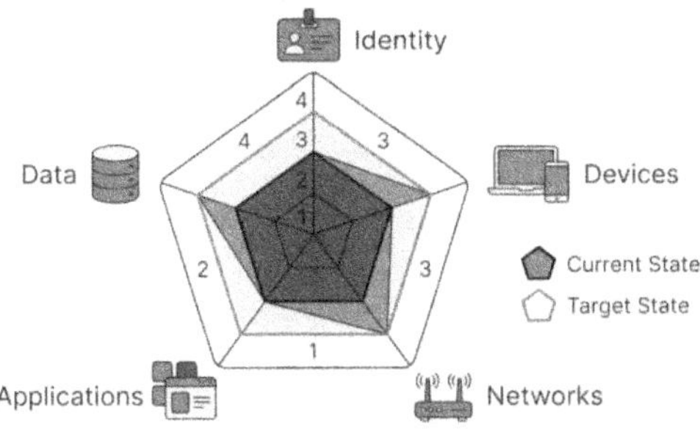

7.3 Piloting Without Paralysis: Running Proof-of-Concept Programs That Inform Production

The pilot phase is where Zero Trust moves from documented intent to operational reality. It is also where many programs lose momentum — either because pilots are designed so carefully that they never surface real friction, or because they are designed so loosely that the findings cannot be generalized to production deployment. A well-designed pilot is neither a demonstration nor a full deployment: it is a structured experiment that produces specific answers to specific questions about how the target controls will perform in the organization's actual operational environment.

The purpose of a pilot is not to validate that the technology works as advertised. By the time a technology reaches the pilot phase, vendor proof-of-concept testing should have established basic functionality. The pilot exists to answer the organizational questions that vendor testing cannot: How does the policy perform in real workflows?

What exceptions does it generate, and are those exceptions acceptable? How do users and support teams respond to the new controls? What operational dependencies were not visible during planning?

7.3.1 Selecting Pilot Populations That Surface Real Friction Points

Pilot population selection is one of the most consequential design decisions in the implementation process. A pilot that runs against a small group of highly technical, change-tolerant users who use only a handful of well-documented applications will produce clean results that do not predict how the policy will perform against the broader workforce. It is the equivalent of testing a medication on the healthiest patients in the trial and concluding that it has no side effects.

Effective pilot populations are chosen to represent the range of access patterns, device types, application dependencies, and workflow complexity that the policy will eventually govern. This typically means including some highly technical users alongside some who are minimally technical, some who use complex multi-application workflows alongside some who use only a few core systems, and some who work entirely on managed devices alongside some who use contractor or BYOD configurations.

The pilot population should also include at least one business-critical workflow — not because the pilot should disrupt it, but because the policy must be validated against it. A conditional access policy that performs well for low-stakes internal tools but breaks the electronic health record workflow or the trading platform will not be deployable in production—discovering that during the pilot, when the scope is limited, and rollback is straightforward, is exactly the right outcome.

Diagram 7.4 – Pilot Population Design Matrix

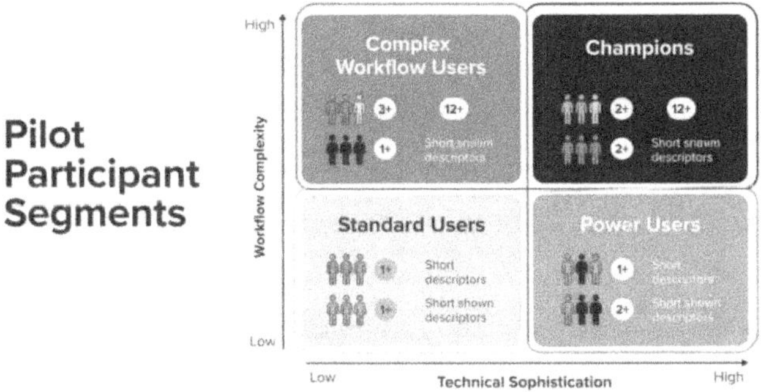

7.3.2 Defining Success Criteria Before Pilots Begin

Success criteria must be defined before the pilot begins, not evaluated after the results are in. Post-hoc success criteria are almost always shaped by what the pilot actually produced, making them useless as decision-making tools. The purpose of pre-defined criteria is to force the program team to commit in advance to what "good enough to proceed" looks like, and to create a basis for honest evaluation when the pilot concludes.

Success criteria for a Zero Trust pilot typically address several dimensions. Policy effectiveness: Does the control enforce what it is designed to enforce, blocking unauthorized access while permitting legitimate access? Exception rate:

What percentage of legitimate access attempts are blocked or challenged, and is that rate acceptable for the organization's operational tempo? Support burden: Does the pilot generate help desk volume that scales to an unacceptable level in production? User experience: Do pilot participants report friction levels that are likely to generate organized resistance during broader deployment?

Each of these criteria should have a specific threshold — not a narrative description of what success would look like, but a measurable value that the program will use to make a binary proceed-or-revise decision. If exception rates above two percent for a given application category indicate a policy that requires revision before production, that threshold should be documented before the pilot launches. The discipline of committing to thresholds in advance is what separates a pilot from a demonstration and makes its findings actionable.

7.4 Metrics That Reflect Security Outcomes, Not Activity Volume

The most common failure mode in Zero Trust reporting is the substitution of activity metrics for outcome metrics. A program that reports the number of endpoints enrolled in device management, the number of users enrolled in MFA, and the number of policies configured is a reporting activity. It is not reporting security outcomes. Activity metrics tell leadership that the program is doing things. Outcome metrics tell leadership whether those things are reducing risk.

The distinction matters because activity metrics are easy to optimize for their own sake and because they can show impressive progress while the security posture remains weak. An organization can have 95% MFA enrollment and still have unprotected service accounts that grant attackers privileged access. It can have a comprehensive microsegmentation policy on paper and still have lateral movement paths that have not been mapped or controlled. Outcome metrics surface these gaps because they measure what is actually happening in the environment, not what policies have been configured.

7.4.1 Leading Indicators: Coverage, Policy Enforcement Rates, and Exception Volumes

Leading indicators are metrics that predict future security posture — they measure the coverage and quality of control enforcement that will eventually reduce the incidence of adverse outcomes. They are most useful in the early and middle phases of a Zero Trust program, when controls are still being deployed, and the program needs signals on whether it is on track to reach its intended coverage.

Policy enforcement rate is one of the most useful leading indicators available. It measures the percentage of access requests to a given resource that are evaluated against the Zero Trust policy, rather than being passed through without evaluation. A rate below 100% means there are access paths that bypass the policy engine entirely — a gap that an attacker can exploit by routing requests through an unenforced channel. Coverage gaps identified through enforcement rate monitoring are directly actionable: they point to specific enforcement points that need to be deployed or configured.

Exception volume is a related indicator that measures how often the policy engine makes access decisions that deviate from the default enforcement posture — either granting access that should be blocked under normal policy or denying access in ways that are subsequently overridden. High exception volumes indicate policy rules that are miscalibrated to the actual workflow environment, and they create a log of decisions that must be reviewed regularly to ensure that exceptions do not accumulate into a de facto alternative access policy.

7.4.2 Lagging Indicators: Mean Time to Detect, Lateral Movement Frequency, Privilege Escalation Rates

Lagging indicators measure security outcomes that have already occurred — they tell leadership whether the controls are actually reducing the incidence and severity of the events they are designed to prevent. They are most useful in the later phases of a Zero Trust program, when controls are sufficiently deployed for their impact on the threat landscape to be measurable.

Mean time to detect measures how quickly the organization identifies a security event from the moment it begins. In a perimeter-centric environment, the median dwell time for a breach — the period between initial compromise and detection — is measured in weeks or months. A Zero Trust program with effective monitoring should significantly reduce dwell time because the continuous verification model generates signals at every access attempt, not just at the perimeter. Tracking this metric over time reveals whether the monitoring investment is actually improving detection speed.

Lateral movement frequency measures how often the security team observes authenticated identities attempting to access resources outside their normal scope. In a well-segmented Zero Trust environment, this metric should be low — not because attackers are not attempting lateral movement, but because the policy engine is blocking unauthorized access attempts and the monitoring system is surfacing the ones that are attempted. A stable or declining lateral movement rate, combined with a stable or increasing detection rate, is one of the clearest signals available that the Zero Trust architecture is functioning as intended.

Privilege escalation rate tracks how frequently identities — human or machine — are observed acquiring permissions beyond their established baseline. This is a particularly important indicator for organizations with complex service account environments and automated pipelines, where privilege creep tends to accumulate silently and create exploitable gaps. A declining escalation rate indicates that least-privilege controls are functioning and that the entitlement review process is reclaiming unnecessary permissions before they can be exploited.

Diagram 7.5 – Zero Trust Metrics Dashboard

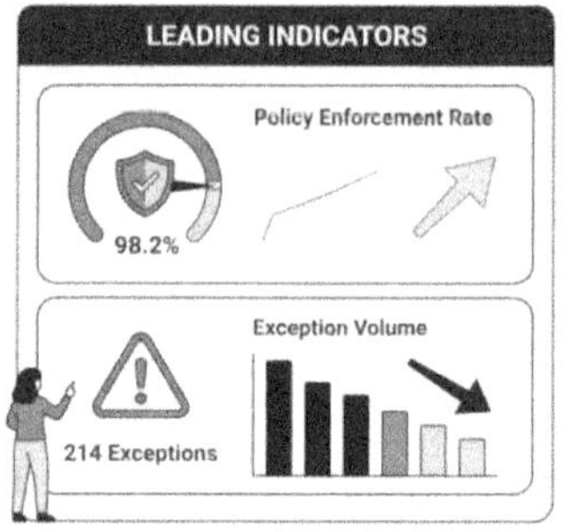

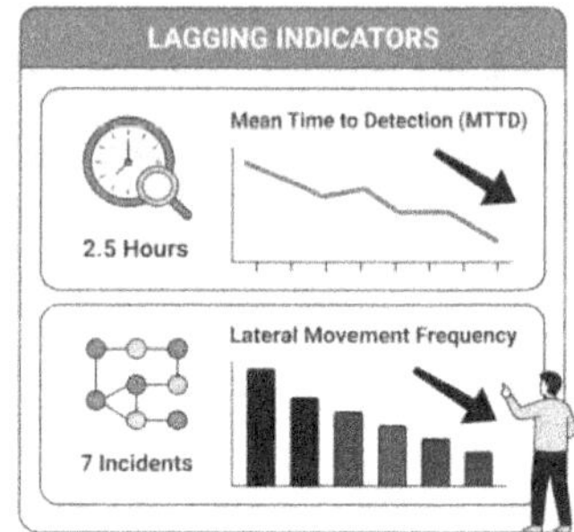

7.4.3 Executive Reporting Formats That Drive Action Rather Than Archive

Executive reporting serves a specific function: it enables senior leaders to make informed decisions about the program. Reporting that is designed to document progress rather than drive decisions tends to accumulate in inboxes without influencing anything. The test of a useful executive report is whether it consistently produces one of three responses: approval to proceed, a request to revise, or a decision to accept a risk — and whether it identifies clearly which of those responses is called for by the current data.

The most effective executive reporting formats for Zero Trust programs are brief, visually structured, and organized around decisions rather than activities. A monthly executive summary that opens with the three decisions leadership needs to make in the next thirty days, followed by the three metrics that are most relevant to those decisions, and closes with the one risk that requires explicit acceptance or escalation, is more useful than a twenty-slide update covering every program activity in the past month.

Formatting choices matter. Trend lines are more informative than point-in-time numbers because they reveal direction. Red-yellow-green status indicators are useful for drawing attention to items that require action. Still, they are credible only when the thresholds that determine the colors are defined in the program's governance documentation rather than assigned retrospectively. Narrative text should be confined to the context and recommendation sections — the data should speak for itself.

Diagram 7.6 – Executive Reporting One-Pager Template

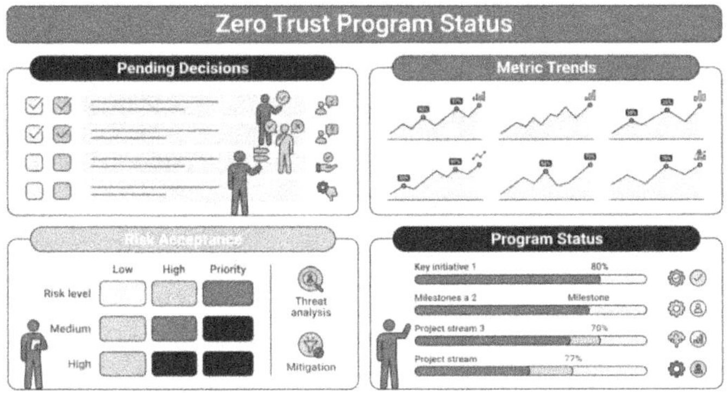

7.5 Manager's Checklist: From Whiteboard to Production

Commission a current-state assessment that maps existing controls against all five Zero Trust capability pillars before sequencing any implementation work.

Produce a capability heat map that shows maturity levels by pillar and identifies the gaps with the highest risk exposure.

Identify at least three quick wins that can be implemented in the first sixty days without creating technical debt or conflicting with the target architecture.

Complete a dependency map before finalizing phase sequencing — identify every prerequisite that must be in place before each subsequent phase can begin.

Set a maturity target for each pillar aligned with organizational risk appetite and regulatory requirements, and document which gaps are accepted and why.

Design pilot populations that include a range of technical sophistication levels, application complexity, device types, and at least one business-critical workflow.

Define pilot success criteria with measurable thresholds before the pilot begins — not after results are available.

Report leading indicators during early phases and lagging indicators once controls reach sufficient coverage — never substitute activity metrics for outcome metrics.

Structure executive reporting around decisions that need to be made, not activities that have been completed.

Review metric thresholds at least quarterly and adjust them as the program matures and the baseline shifts.

7.6 What the Roadmap Actually Measures

A Zero Trust roadmap that cannot tell decision-makers whether security posture is improving is not a roadmap; it is a schedule. The discipline that separates successful implementations from stalled ones is not technical sophistication but measurement rigor. Organizations that define what progress looks like before the first control is deployed, that build their phase sequence around dependencies rather than deadlines, and that translate maturity gaps into risk language rather than technology language create the conditions for a transformation that survives budget cycles and leadership transitions.

The CISA maturity model, properly applied, gives the program a shared vocabulary. The dependency map gives it a realistic sequence. The pilot framework gives it a mechanism for learning without committing the full environment. The metrics framework gives leadership the signals it needs to govern without micromanaging. Together, these tools turn a whiteboard vision into a navigable program, one that can demonstrate its value incrementally, absorb the inevitable surprises, and sustain itself through the multi-year arc of a genuine architectural transformation.

The chapter that follows addresses the organizational dimension of that transformation: the people, teams, and cultural dynamics that will determine whether the architecture survives contact with operational reality.

8 The Human Architecture: Organizational Change, Team Design, and Cultural Transformation

Six months into the implementation of Zero Trust at a large regional bank, the program manager received a call from the head of retail operations. Branch managers across four states were informally bypassing the new conditional access policy by sharing a service account whose credentials had been approved to avoid the multi-factor prompt. The workaround had started with one frustrated branch, spread through an unofficial managers' group chat, and was now standard practice for a workforce of roughly eight hundred people. The technology was working exactly as designed. The change management had not worked at all.

This scenario is not exceptional — versions of it appear in nearly every large-scale Zero Trust deployment. The access policy that seemed straightforward in the architecture review generates unexpected friction in a specific operational context. Users who were not consulted during design find workarounds faster than the security team can discover and close them. The result is a two-tier environment: the architecture documented and the access patterns that govern daily operations. Closing that gap is not a technology problem. It is a people-and-organizational design problem.

Why it matters: the tools and capabilities described in the preceding chapters — policy engines, microsegmentation, behavioral analytics — only reduce risk if they operate in an environment where people understand what they are doing and why, where teams have the roles and skills needed to manage them, and where the organizational culture treats security behaviors as a genuine professional responsibility rather than a compliance burden. An architectural transformation that neglects these dimensions will produce a security posture on paper and a different, weaker posture in practice. The human architecture must be designed with the same intentionality as the technical one.

This chapter addresses the organizational and cultural dimensions of a Zero Trust transformation. It covers how to map the stakeholder landscape before change begins, how to restructure security teams around the functional roles that Zero Trust demands, how to apply change management disciplines that are calibrated for the specific dynamics of security transformation, and how to build a workforce culture in which security norms are internalized rather than merely performed. Throughout, the manager's lens is the same as in any other chapter: what decisions does this require, what risks does it address, and what actions does it support?

8.1 Mapping the Human Stakeholder Landscape Before Change Begins

Every organizational change has a stakeholder landscape — a population of people whose roles, workflows, or authority are affected by the change and who will respond to it in ways that either accelerate or impede progress. In a Zero Trust transformation, that landscape is unusually wide. It extends from the executive team that must fund and champion the program through the engineering teams that must implement it, the IT operations staff whose daily procedures will change, the business unit leaders whose workflows will be affected by new controls, and the end users who will encounter new authentication prompts and access restrictions in their daily work.

Mapping this landscape is not a sociological exercise — it is a matter of risk management. Stakeholders who are not identified in advance, whose concerns are not addressed during program design, and whose workflows are disrupted without adequate preparation become sources of implementation resistance that can delay or derail phases. The branch bank scenario described at the opening of this chapter was the product of a stakeholder mapping failure: the retail operations workforce was treated as an end-state recipient of the change rather than a participant in its design.

8.1.1 Identifying Champions, Skeptics, and the Uninformed Majority

Stakeholders in any organizational transformation sort into three broad categories. Champions are individuals or teams who understand the rationale for the change, can articulate its benefits in terms relevant to their context, and are willing to advocate for it within their networks. Skeptics have concerns — about operational disruption, resource demands, and the program's feasibility — that are often legitimate but can harden into active opposition if they are not engaged constructively. The uninformed majority neither supports nor opposes the program because they do not yet understand what it means for them.

Champions should be identified early and engaged as co-designers wherever possible. A network engineer who has been frustrated by the limitations of the perimeter model for years, a compliance officer who sees Zero Trust as a path to cleaner audit findings, a business unit leader whose team suffered a breach enabled by lateral movement — these individuals bring credibility that the security team alone cannot provide when the program encounters resistance.

Skeptics deserve careful attention rather than dismissal. Their concerns frequently point to real gaps in the program design — workflows that have not been mapped, dependencies that have not been accounted for, operational constraints that will require policy exceptions. Engaging skeptics through structured listening sessions and visibly incorporating their input into the program design converts a potential source of resistance into a source of quality control. Skeptics who are heard and whose concerns are reflected in program decisions often become the program's most credible validators.

The uninformed majority is reached through sustained communication — not a single all-hands announcement, but a structured cadence of messages that provide context, set expectations, and answer questions before the changes arrive. The communication plan is discussed in more detail in the change management section of this chapter.

8.1.2 Executive Sponsorship: What Leaders Must Actually Do, Not Just Endorse

Executive sponsorship is cited as a success factor in nearly every organizational change methodology, and Zero Trust is no exception. But the quality of sponsorship varies enormously, and the most common form — a vice president or CIO who sends an endorsement email at the program launch and then delegates all subsequent engagement to the security team — is also the least useful.

Effective executive sponsorship for a Zero Trust program requires three things that endorsement alone does not provide. First, decision authority: sponsors must be empowered to resolve escalations that cannot be handled at the program level, including conflicts between security requirements and operational preferences from senior business unit leaders. Second, resource commitment: sponsors must protect the program's staffing, budget, and timeline against the competing priorities that emerge in every budget cycle. Third, behavioral modeling: sponsors who visibly comply with the same authentication requirements and access controls as the rest of the workforce signal that the program is not a security team initiative but an organizational commitment.

The manager's role is to support sponsors by making escalation easy and decision-making efficient. A sponsor who is asked to arbitrate every minor policy conflict will disengage. A sponsor who receives a monthly one-page summary of the two or three decisions requiring their authority, with a clear recommendation and the information needed to evaluate it, will remain engaged and effective throughout the program.

Diagram 8.1 – Stakeholder Landscape Map for Zero Trust Transformation

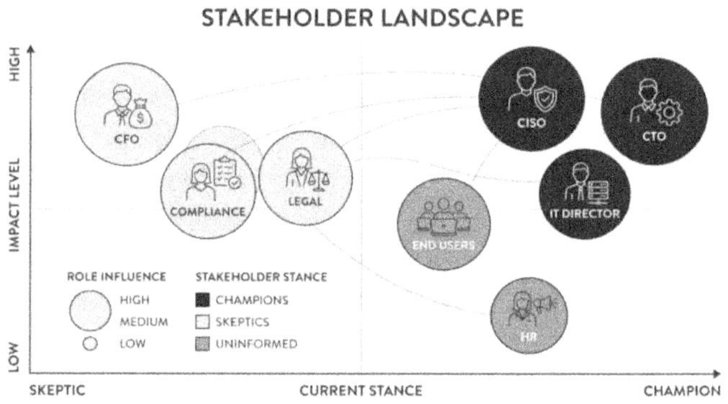

8.1.3 Cross-Functional Coalition Building: IT, Legal, HR, and Business Units

Zero Trust affects organizational functions well beyond the security team, and its successful implementation requires active participation from several of them. Legal and compliance teams need to understand how Zero Trust controls interact with data protection obligations, e-discovery requirements, and regulatory reporting. Human resources must be involved when the program changes the scope of employee monitoring, introduces new authentication requirements for sensitive HR systems, or requires updates to acceptable use policies. Business unit leaders must participate in designing the access policies that govern their workflows.

Building a cross-functional coalition is not the same as sending a stakeholder notification. It means establishing a governance structure in which representatives from these functions have genuine input into program design, a defined process for raising and resolving concerns, and visibility into the program's timeline that allows them to plan their own participation. The security team's instinct to limit the coalition to people who can move at the security team's pace is understandable but counterproductive — the functions that move more slowly are often the ones whose concerns, if unaddressed, create the most significant implementation obstacles.

8.2 Restructuring Security Teams Around Zero Trust Functional Roles

The organizational structure of most enterprise security teams reflects the perimeter security model they were built to maintain. There is a network security function that manages firewalls and intrusion detection at the boundary. There is an identity and access management function, often partially embedded in IT operations, that manages directories and authentication systems. There is a security operations center that monitors alerts from perimeter and endpoint tools. And there is an application security function that may or may not be integrated with the rest of the team.

This structure was designed for an environment where the primary security work occurred at the network boundary, identity management was a relatively static administrative function, and monitoring focused on detecting threats that had already crossed the perimeter. Zero Trust changes all three of these assumptions simultaneously. The boundary is no longer the primary enforcement point. Identity management becomes a continuous, dynamic, policy-driven discipline. Monitoring must cover every access interaction across every layer of the environment, not just the perimeter alerts.

Restructuring the security team for Zero Trust does not require a complete organizational rebuild, and in most cases, it should not. It requires adding new functional roles, redefining existing ones, and creating structural connections between functions that have traditionally operated in relative isolation. The three roles that most commonly need to be created or significantly strengthened are the identity and access engineering function, the policy operations function, and the integration between security engineering and security operations.

8.2.1 The Identity and Access Engineering Function

In a Zero Trust architecture, identity governance is not a configuration task — it is an ongoing engineering discipline. The policies that govern who gets access to what, and under what conditions, must be continuously updated to reflect changes in the workforce, the application portfolio, the threat environment, and the regulatory landscape. The systems that enforce those policies — identity providers, policy engines, conditional access platforms — must be maintained, integrated, and monitored with the same rigor applied to any other critical infrastructure component.

The identity and access engineering function is responsible for this discipline. Its scope includes the technical design and maintenance of identity infrastructure, the development and testing of access policies, the management of the identity lifecycle from provisioning through deprovisioning, and the integration of identity signals into the broader Zero Trust policy engine. It also includes the governance of non-human identities — service accounts, API keys, machine certificates, and increasingly the AI agent identities that are proliferating across enterprise automation pipelines.

This function requires skills that overlap between traditional IAM engineering, cloud platform engineering, and security policy design. Most organizations do not have a single team with this combination of competencies, which means the restructuring process must include a skills assessment that identifies gaps and a plan for addressing them through hiring, training, or strategic vendor partnerships. The manager who builds this function correctly creates the organizational capability that supports every other Zero Trust capability layer.

8.2.2 Policy Operations: The Team That Keeps Controls Current

A Zero Trust policy engine is only as effective as the policies it enforces. Policies that are well-calibrated at deployment will drift out of alignment with operational reality as applications change, workforces evolve, and threat patterns shift. Without a dedicated function responsible for maintaining policy accuracy, exception management, and periodic review, the policy engine becomes a source of unnecessary access friction rather than a precision enforcement mechanism.

The policy operations function — sometimes called the Zero Trust policy office or access governance team — is responsible for the ongoing management of the access policies that govern the Zero Trust environment. Its work includes reviewing and processing access exceptions, conducting periodic entitlement reviews to identify over-provisioned accounts, updating policies to reflect organizational changes such as workforce restructuring or application migrations, and analyzing exception and access denial logs to identify policy calibration issues.

This function occupies a critical coordination role in the Zero Trust program because it sits at the intersection of security requirements, operational needs, and business workflows. The policy operations team is the organizational mechanism through which the Zero Trust architecture remains responsive to operational reality without compromising security principles. It is also the team that is most visible to end users when something does not work — which means its staffing, responsiveness, and communication quality directly affect the workforce's experience of the Zero Trust program.

Diagram 8.2 – Zero Trust Security Team Structure

Zero Trust Security Team Structure

8.2.3 Bridging the Gap Between Security Engineering and Security Operations

Security engineering and security operations are organizationally distinct in most enterprises, and they are often culturally distinct as well. Engineering teams design and build controls; operations teams monitor and respond to events. In a perimeter-centric environment, these functions interact primarily when a new perimeter control needs to be integrated into the monitoring stack. In a Zero Trust environment, the interaction is continuous and bidirectional: monitoring findings must feed back into policy design, and policy changes must be communicated to the operations team so that alert thresholds can be adjusted accordingly.

Bridging this gap requires structural mechanisms rather than cultural aspirations. A shared backlog that includes both engineering tasks and monitoring-driven remediation items creates operational visibility across the boundary. A regular forum in which the operations team presents detection findings and the engineering team presents upcoming policy changes gives each side the context it needs to do its work effectively. Shared metrics that measure outcomes — reduction in lateral movement, improvement in detection time — rather than function-specific outputs create aligned incentives.

Organizations that invest in this bridge build a feedback loop that is one of the most powerful capabilities a Zero Trust program can have: the ability to translate what the monitoring environment is seeing directly into policy adjustments that close the gaps it reveals. This loop is what transforms a Zero Trust deployment from a static configuration into a continuously adaptive security program.

8.3 Change Management Techniques Calibrated for Security Transformations

Change management in security transformations has a specific character that distinguishes it from change management in other organizational initiatives. The changes are often invisible to end users until they encounter friction — a new authentication prompt, a blocked application, an access request that now requires approval. The benefits are largely defined by events that do not happen — breaches that are prevented, lateral movement that is blocked. And the timeline is long enough that the initial momentum from launch must be sustained through years of incremental implementation.

Standard change management frameworks provide a useful starting point, but they must be adapted for these dynamics. The communication plan that works for an ERP system migration, where the change is visible and the benefits are tangible, does not translate directly to a Zero Trust transformation where the security improvements are real but largely invisible. The engagement approach that works for a voluntary digital adoption program does not translate to a mandatory security control rollout where non-compliance creates organizational risk.

8.3.1 Communicating the "Why" in Terms That Resonate with Non-Security Audiences

The most common communication failure in security transformations is explaining the "what" and skipping the "why." Security teams know why Zero Trust matters — the threat landscape is clear to them, the failure modes of the perimeter model are well understood, and the risk calculus is familiar. For the retail operations manager, the clinical nurse, or the financial analyst, none of that context is self-evident. They encounter a new authentication prompt and experience it as friction. Without a compelling "why," friction is the entire story.

Effective "why" communication is grounded in scenarios that are relevant to the specific audience's context. For a healthcare organization, the "why" might be framed around protecting patient records from the kind of breach that has disrupted hospitals across the country — a scenario that clinical staff understand intuitively and care about deeply. For a financial institution, it might be framed around protecting client account data from the credential theft attacks that have compromised competitors. For a government agency, it might be tied directly to the regulatory mandates that make Zero Trust not just a good practice but a compliance requirement.

The communication plan should be organized around audiences rather than messages. Different stakeholder groups need different framings of the same underlying message. Executive communications should focus on risk reduction and program governance. Business unit communications should focus on workflow impact and the support resources available when friction occurs. End user communications should focus on what is changing, why it matters, and what to do when the new controls create a problem. Each of these audiences requires its own communication channel, cadence, and format.

Diagram 8.3 – Change Communication Architecture

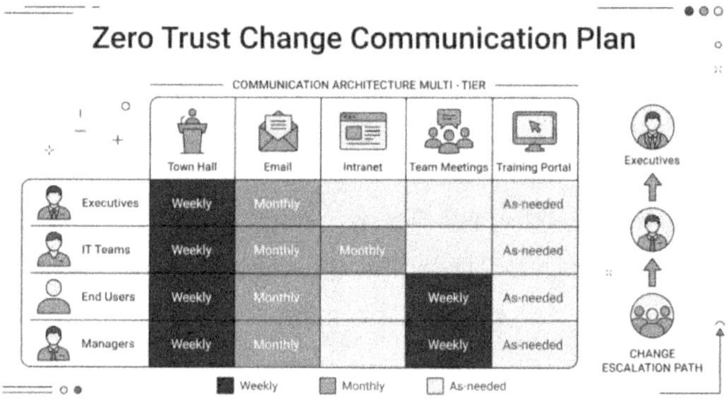

	Town Hall	Email	Intranet	Team Meetings	Training Portal
Executives	Weekly	Monthly			As-needed
IT Teams	Weekly	Monthly	Monthly		As-needed
End Users	Weekly	Monthly		Weekly	As-needed
Managers	Weekly	Monthly		Weekly	As-needed

8.3.2 Sequencing User-Facing Changes to Minimize Productivity Disruption

The sequencing of user-facing changes is a change management decision with direct security implications. Rolling out all new authentication requirements simultaneously maximizes disruption and help desk load, increases the likelihood of workarounds, and creates a large body of user experience problems that must be diagnosed in parallel. Sequencing changes progressively — starting with the user populations and application contexts where friction is lowest, using the learning from those deployments to refine the approach before moving to higher-friction contexts — reduces all of these risks.

A combination of risk priority and operational impact should drive sequencing. The access controls for the most sensitive systems — privileged administrative accounts, financial transaction systems, patient record applications — should be implemented first, as that is where the greatest risk reduction is. But implementation in those contexts should be preceded by thorough workflow mapping, a well-resourced support plan, and a clear escalation process for the access issues that will inevitably arise in complex operational environments.

The timing of user-facing changes relative to operational peaks matters as well. A healthcare organization that deploys new authentication requirements at the beginning of a flu season, when clinical staff are operating at maximum capacity and tolerance for friction is lowest, will generate more resistance and more workarounds than one that deploys in a lower-volume period and provides a structured onboarding experience. This kind of operational timing awareness requires the program team to engage with business unit calendars rather than planning deployments solely on the basis of technical readiness.

8.4 Building a Security-Aware Workforce Without Manufacturing Compliance Theater

Security awareness programs have a credibility problem in many organizations. They are associated with annual training modules that test knowledge of phishing indicators, compliance attestation exercises that employees click through in under five minutes, and awareness campaigns that produce measurable behavior change in the week they run and negligible change in the months that follow. This association is not unfair — most security awareness programs are designed to produce compliance evidence rather than behavioral change.

Zero Trust creates an opportunity to rethink this entirely. Because Zero Trust controls are present in the daily workflow — in the authentication experience, in the access decisions that the policy engine makes, in the alerts that are generated when something unusual occurs — they create natural touchpoints for security education that are grounded in operational reality rather than abstract scenarios. The goal is to use those touchpoints to build genuine understanding rather than procedural compliance.

8.4.1 Training That Changes Behavior Rather Than Tests Knowledge

Behavioral security training is grounded in a principle that most knowledge-testing approaches ignore: knowing the right answer to a multiple-choice question about phishing does not predict how someone will behave when they encounter a convincing social engineering attempt under deadline pressure. Behavior change requires practice in context, corrective feedback close to the moment of the behavior, and reinforcement over time. Knowledge testing provides none of these.

Training programs that change behavior are designed around the specific decisions that employees actually face in their work. For most knowledge workers, those decisions center on authentication — responding to MFA prompts correctly, recognizing suspicious authentication requests, understanding why access to a system they use regularly might be blocked under certain conditions. Simulated exercises that test these decisions in realistic operational contexts, with immediate feedback and clear explanation, produce measurable changes in response behavior over time.

The manager's role in training program design is to ensure that the scenarios reflect operational reality and that the training cadence is appropriate for the workforce's pace. A training program that runs quarterly modules covering abstract scenarios produces different outcomes than one that delivers brief, contextual micro-trainings at the point where the relevant decision occurs — for example, a thirty-second explainer that appears when a user first encounters a new authentication requirement, explaining what it is and why it matters.

Diagram 8.4 – Behavioral Training Design Framework

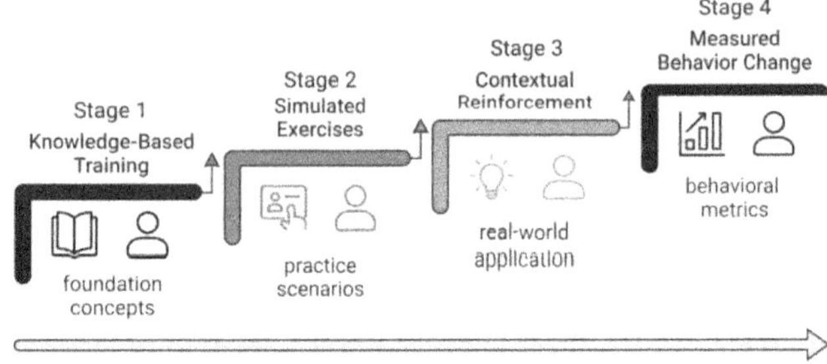

8.4.2 Reinforcing Zero Trust Norms Through Leadership Visibility

Leadership behavior is one of the most powerful signals available to an organization about what is genuinely expected versus what is merely stated in policy. When a senior executive uses a personal device to bypass the device registration requirement, or requests a standing exception to the privileged access policy because multi-factor prompts are inconvenient, that behavior communicates something that no security awareness training can counteract. It communicates that the security requirements are negotiable for people with sufficient authority, and that the stated norm is not the real one.

Conversely, when leaders visibly comply with the same controls as the rest of the workforce — when the CISO mentions in a team meeting that she received and correctly responded to a suspicious authentication prompt, or when a business unit vice president publicly declines to request an exception he could have obtained — those behaviors communicate that the security norms are organizational rather than bureaucratic. They normalize the expected behavior in a way that policy documents and training modules cannot.

Building leadership visibility into the Zero Trust change management plan means explicitly asking senior leaders to behave in certain ways, providing them with talking points that explain those behaviors in terms their teams will find credible, and recognizing publicly when leadership behavior supports the program's goals. This is not manipulation — it is the application of organizational behavior principles that are well-established in change management research and that security programs have historically underutilized.

8.4.3 Measuring Cultural Shift: Surveys, Incident Patterns, and Help Desk Signals

Cultural shift is difficult to measure directly, but it produces observable signals that can be tracked over time. Security programs that are building genuine cultural change see those changes reflected in employee survey responses, in the frequency and nature of security incidents, and in the patterns of help desk contacts that the new controls generate.

Employee surveys can measure security culture dimensions that operational metrics cannot capture, such as whether employees understand why the controls exist, whether they view the security team as a resource rather than an obstacle, and whether they are willing to report suspicious activity. Surveys administered before a major change phase and again six to twelve months afterward provide a baseline comparison that reveals whether the change management investment is producing cultural movement or simply compliance activity.

Incident patterns reveal cultural health in a different dimension. An organization with a healthy security culture generates a high volume of employee-reported incidents because employees observe their environment, recognize suspicious behavior, and report it through the appropriate channels. An organization whose incident reports are almost entirely system-generated, with minimal employee contribution, has a workforce trained to be compliant rather than engaged. The ratio of employee-reported to system-detected incidents is a more sensitive indicator of cultural engagement than any survey question.

Help desk signals reveal the friction points that change management did not adequately address. A high volume of contacts regarding a specific authentication requirement in a specific business unit indicates either a workflow design problem, a communication gap, or both. Help desk trend data, reviewed regularly by the program team, serves as an early warning system for organizational friction that, if left unaddressed, generates workarounds and erodes the security posture the controls are designed to maintain.

Diagram 8.5 – Cultural Shift Measurement Dashboard

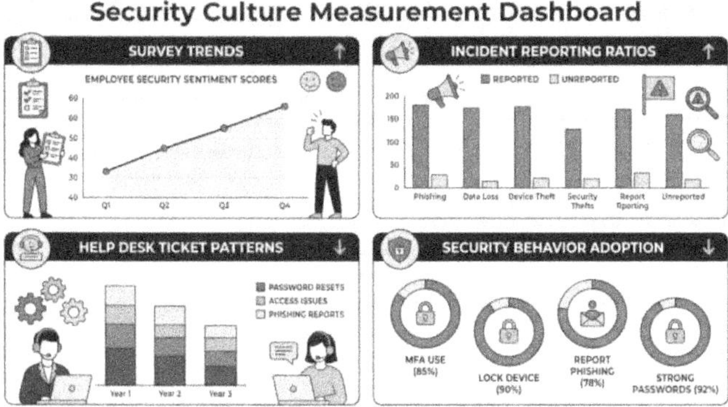

8.5 Manager's Checklist: The Human Architecture

Complete a stakeholder mapping exercise before the first phase of deployment — identify champions, skeptics, and the uninformed majority across all affected organizational functions.

Establish executive sponsorship that includes decision authority, resource protection, and behavioral modeling — not just endorsement communications.

Build a cross-functional coalition with legal, HR, and business-unit representation, and establish a governance mechanism for raising and resolving concerns.

Create or strengthen the identity and access engineering function with skills that span IAM, cloud platform, and security policy domains.

Establish a policy operations function responsible for exception management, entitlement reviews, and ongoing policy calibration.

Create structural mechanisms — shared backlogs, joint forums, shared metrics — that bridge the gap between security engineering and security operations.

Develop a communication plan organized by audience, not by message — each stakeholder group needs a tailored framing of the same core narrative.

Sequence user-facing changes based on risk priority and operational impact, and coordinate deployment timing with business unit calendars.

Design training programs around behavioral change rather than knowledge testing — use contextual micro-trainings at workflow touchpoints rather than periodic awareness modules.

Measure cultural shift through a combination of employee surveys, incident pattern analysis, and help desk signal review — not through training completion rates alone.

Track exception request volume and workaround patterns as leading indicators of change-management failure before they become security-posture failures.

8.6 The Architecture That People Build

The most technically sophisticated Zero Trust architecture in the organization's documentation does not reduce risk on its own. It reduces risk when the people responsible for operating it have the skills and team structures to maintain it, when the people affected by it understand what it is asking of them and why, and when the organizational culture treats the behaviors it requires as professional norms rather than bureaucratic obligations. These conditions do not emerge from technological deployment. They are built through deliberate organizational design — stakeholder engagement, role development, communication discipline, and cultural reinforcement over time.

The manager who treats human architecture with the same intentionality as the technical architecture creates the conditions for a program that works in operational reality, not just in vendor demonstrations and architecture reviews. The program that neglects the human dimension will find that its controls are technically sound but operationally circumvented — a situation that is in some ways more dangerous than having no controls at all, because it produces false confidence in a posture that is actually permeable.

The next chapter addresses how the program sustains itself once the initial deployment energy fades: the monitoring capabilities, governance structures, and disciplines of continuous adaptation that keep a Zero Trust program effective over the years and through the organizational transitions that follow the first production deployment.

9 The Program That Never Finishes: Continuous Monitoring, Adaptive Governance, and Long-Term Resilience

Eighteen months after a regional health system completed the first phase of its Zero Trust deployment, a new CISO arrived from the private sector. She requested a briefing on the program's current state. What she received was a comprehensive slide deck documenting the architecture as it had been designed at deployment: the identity policies, the segment boundaries, the enforcement points, and the monitoring coverage. What the briefing did not contain was any information about what had happened since deployment. Three data centers had been migrated to a cloud provider. A large practice management group had been acquired and partially integrated. A critical application had been replaced with a SaaS platform that the original network segmentation policy did not account for. The architecture in the slide deck had not changed. The actual environment had changed substantially.

This is the most common failure mode in mature Zero Trust programs — not in the initial deployment, but in maintaining the program's alignment with an environment that never stops changing. Applications are replaced. Acquisitions add new network segments and identity populations. Regulatory requirements evolve. New threat patterns emerge. Each of these changes creates a potential gap between the Zero Trust controls as designed and those as enforced. Without continuous discipline of monitoring, review, and adaptation, those gaps accumulate until the architecture is a historical document rather than a living security posture.

Why it matters: the security posture of a Zero Trust environment degrades not through dramatic failures but through the steady accumulation of small misalignments. An enforcement point that is not updated when a new application is deployed. A policy that governs a workflow that no longer exists. A behavioral baseline that has not been recalibrated since the workforce composition changed. Each of these is a manageable problem when caught early. Together, they constitute a posture that appears intact in the architecture diagram but is demonstrably weaker in the operational environment. Continuous monitoring and adaptive governance are the disciplines that prevent this accumulation.

This chapter addresses the monitoring capabilities, behavioral analytics practices, governance structures, and horizon planning disciplines that sustain a Zero Trust program over the long arc of organizational change. It is, in a sense, the most important chapter in the book for managers who have already begun their Zero Trust journey — because the decisions that determine whether a program remains effective or slowly atrophies are governance and operational, not architectural.

9.1 Telemetry Architecture for a Zero Trust Environment: Seeing Everything, Storing What Matters

Zero Trust generates more telemetry than perimeter-based security, because it evaluates every access request rather than only the traffic that crosses the network boundary. Every policy decision — approved, denied, or challenged — produces a log entry. Every device posture assessment produces a signal. Every identity authentication generates an event. In a large enterprise, this data volume is substantial, and managing it effectively requires architectural thought rather than simply deploying more storage.

The telemetry architecture question is not "how do we collect everything" but "how do we collect the right things, correlate them effectively, and store them for the right duration." These are three distinct design challenges, and organizations that conflate them tend to build monitoring environments that are simultaneously expensive and incomplete — generating enormous volumes of data while missing the signals that would reveal an active threat.

9.1.1 Log Sources, Coverage Gaps, and the Cost of Visibility

The first telemetry architecture task is inventory. What log sources are currently feeding the environment? What events do they capture, and at what level? Where are the coverage gaps — the access paths, devices, or environments from which no signals are currently collected? This inventory exercise often reveals surprises: cloud workloads that are generating access events with no corresponding log collection, legacy systems whose logging capabilities have never been configured, and network segments that are effectively invisible to the monitoring environment.

Coverage gaps in a Zero Trust monitoring environment are operationally equivalent to enforcement gaps: they represent access paths where adversary activity can occur without generating a detectable signal. The monitoring coverage map should be maintained alongside the enforcement point map as a joint artifact that reveals the relationship between what is being controlled and what is being seen. A control that is enforced but not monitored provides access governance without detection capability — it will block unauthorized access attempts. Still, it will not surface the access patterns that indicate an active threat.

The cost of visibility is real and must be managed explicitly. Log collection, transport, processing, and storage all consume resources, and the costs scale with the volume and fidelity of the data collected. The discipline of cost management in telemetry architecture is to differentiate between log sources based on their signal value — the probability that their data will contribute to a detection finding — and to collect high-signal sources at full fidelity while reducing collection frequency or fidelity for low-signal sources. This requires an ongoing assessment of which sources have contributed to detections and which have generated volume without an actionable signal.

Practical telemetry prioritization follows a signal value hierarchy. Identity provider authentication logs, privileged access session recordings, and policy engine decision logs are high-signal sources that should be collected at full fidelity with minimal filtering. General network flow data from well-segmented, low-risk environments may be collected at a lower sampling rate without significant loss of detection. Application performance logs from non-sensitive systems may be excluded from the security telemetry pipeline entirely while remaining available to operations teams through separate channels. The decisions about what to collect, at what fidelity, and for how long governance decisions are that require explicit ownership and regular review.

Diagram 9.1 – Zero Trust Telemetry Coverage Map

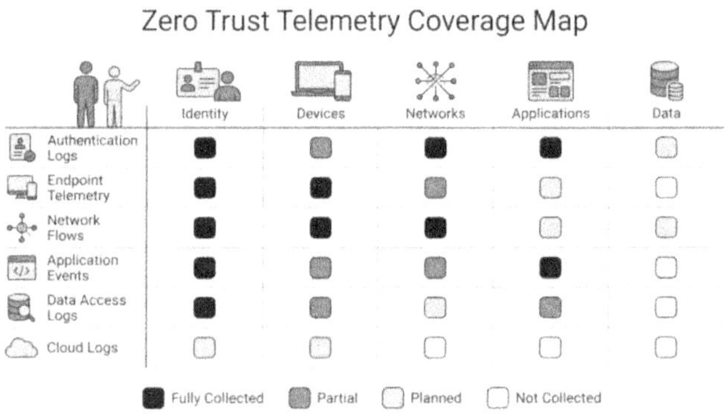

9.1.2 Correlating Identity, Device, and Network Signals Into Coherent Threat Narratives

Individual log sources tell partial stories. An authentication event from an unusual geographic location is a weak signal on its own — it might indicate a traveling employee, a VPN user, or an attacker. The same authentication event, combined with a device posture assessment showing that the endpoint failed its security checks in the past hour, and a network signal indicating that the session is accessing a data repository the user has not touched in six months, tells a much more coherent story. The correlation of signals across the identity, device, and network layers is what transforms raw telemetry into actionable threat intelligence.

Building effective correlation capability requires three things. First, a common data model — a consistent schema that allows events from identity providers, device management platforms, and network monitoring tools to be joined on shared attributes such as user identity, device identifier, and timestamp. Without a common schema, correlation requires manual data transformation, which does not scale to the volume of events generated by a Zero Trust environment.

Second, correlation rules or models that reflect the specific threat scenarios the organization is most exposed to. Generic correlation rules produce generic detections. An organization whose primary threat is insider data exfiltration needs correlation logic that surfaces patterns of unusual data access at unusual times by authenticated users. An organization whose primary threat is credential theft and subsequent lateral movement needs correlation logic that surfaces authentication from new devices, followed by access to sensitive resources. Both organizations need correlation capability, but the specific logic differs.

Third, sufficient computational infrastructure to run correlation in near-real time against the full volume of telemetry being collected. A correlation that runs on a twenty-four-hour batch cycle is not suitable for detecting active threats — by the time the correlation identifies the attack pattern, the lateral movement has already propagated, and the data exfiltration has already occurred. The investment in real-time correlation infrastructure is therefore directly tied to the mean time-to-detect metric introduced in the previous chapter.

9.1.3 Retention Policies Calibrated to Detection Timelines and Regulatory Requirements

Retention policies determine how long telemetry data is kept and in what form it is kept. They must be calibrated against two competing requirements: the detection timeline — how far back investigation of a detected incident might need to look — and the cost of storage, which scales with both the volume of data and the duration of retention.

Detection timelines vary significantly by threat type. Lateral movement by a credential-theft attacker may be detected within hours or days of the initial authentication event. Insider threat activity, conducted slowly by a user who understands the monitoring environment, may not be detected until behavioral patterns have accumulated over weeks or months. Regulatory requirements add a third dimension: some sectors require log retention for specific periods regardless of detection timelines, and forensic investigation requirements after a confirmed breach may demand access to data from months before the breach was detected.

A practical retention architecture typically employs tiered storage: high-fidelity, immediately accessible storage for recent data — typically thirty to ninety days — combined with compressed, lower-cost archival storage for older data that can be retrieved for investigation when needed. The boundary between tiers should be set based on the organization's detection timeline experience rather than on storage cost alone. An organization that regularly detects threats within fourteen days of initial activity can set a shorter immediate-access window than one whose threat patterns emerge over longer periods.

Diagram 9.2 – Telemetry Retention Architecture

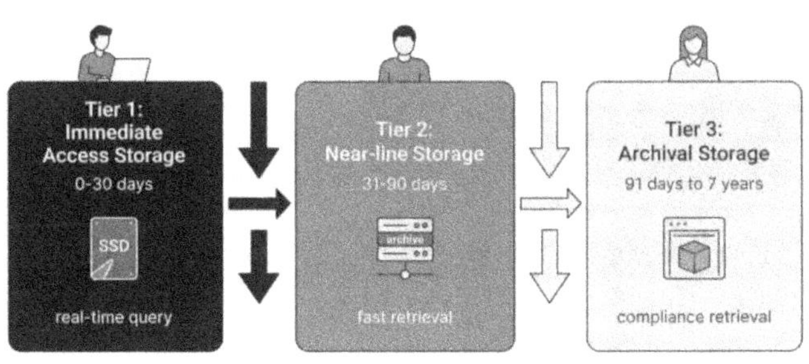

9.2 Behavioral Analytics and Anomaly Detection as Ongoing Enforcement

Zero Trust policy engines make access decisions based on the signals available at the moment of an access request. They are, by design, present-tense instruments — they evaluate the current posture of the identity, device, and context and render a decision accordingly. What they cannot do, in isolation, is detect threats that develop over time through a series of individually authorized access events.

The attacker who uses a legitimately provisioned credential to access a series of sensitive resources, each of which is individually authorized, will not be blocked by a policy engine that evaluates each request in isolation.

Behavioral analytics addresses this temporal gap. By analyzing patterns of access behavior over time and comparing them to established baselines, behavioral analytics systems can identify sequences of authorized actions that collectively suggest adversarial intent — even when no individual action would have triggered a policy denial. This capability closes the gap between Zero Trust's access governance function and the threat-detection function that the monitoring environment must provide.

9.2.1 Baselining Normal to Surface Abnormal Without Alert Fatigue

The fundamental challenge of behavioral analytics is defining what "normal" means for a specific identity or entity in a given operational context. A baseline that is too narrow will generate excessive false positives. Every deviation from the tightest historical pattern will be flagged, producing an alert volume that the operations team cannot process effectively. A baseline that is too broad will generate excessive false negatives. This anomalous behavior indicates that a threat will fall within the wide bounds of "normal" and generate no alert at all.

Effective baseline construction requires enough historical data to capture the natural variation in an identity's access patterns across different operational conditions: day versus night, weekday versus weekend, high-workload periods versus routine periods, on-premises access versus remote access. It also requires awareness of the organizational context that shapes those patterns — a user whose role involves quarterly auditing of systems they do not access otherwise will appear anomalous every quarter unless the baseline reflects the cyclical nature of their access patterns.

Baseline models must be updated continuously as access patterns legitimately evolve. A user who is promoted, transferred to a new role, or assigned to a new project will have different access patterns after the transition than before it. A behavioral analytics system that does not incorporate legitimate pattern changes will generate a persistent stream of low-confidence alerts about normal activity in new contexts — exactly the kind of alert noise that erodes operations team confidence in the system's signal quality and leads to systematic alert suppression.

Alert fatigue is one of the most serious operational risks in behavioral analytics. When the alert volume exceeds the operations team's capacity to investigate, triage discipline degrades. Teams begin to suppress alerts based on volume rather than content, and the high-confidence detections that require immediate response get lost in the noise generated by miscalibrated baselines. Managing alert fatigue requires ongoing investment in baseline quality, alert-prioritization logic, and honest feedback loops between the operations and analytics engineering teams.

9.2.2 User and Entity Behavior Analytics in a Zero Trust Context

User and Entity Behavior Analytics — commonly referred to as UEBA — applies behavioral analysis not just to human user identities but to the full range of entities in the Zero Trust environment: service accounts, applications, cloud workloads, and increasingly AI agents. In a Zero Trust architecture where the principle of least privilege applies equally to human and non-human identities, behavioral anomalies in service account activity or application communication patterns can be as significant as anomalies in human user behavior.

Service account behavior is particularly important to monitor because service accounts tend to exhibit highly consistent, automated access patterns, making deviations from the baseline unusually significant. A service account that accesses the same three endpoints on a consistent schedule for months and then begins accessing a new endpoint, or begins running at an unusual time, has deviated from a baseline that is specific enough to make the deviation meaningful with high confidence. These deviations are frequently associated with compromise of the service account credential or modification of the automation process that drives the account's activity.

Application and workload behavior analytics apply the same principles to the communication patterns between system components. A microservice that consistently communicates with a small set of downstream dependencies and then begins communicating with an unfamiliar endpoint is exhibiting the behavioral signature of several attack patterns — including supply chain compromise, configuration manipulation, and unauthorized data exfiltration via a compromised workload. Monitoring this layer of entity behavior requires instrumentation at the application communication level, which in cloud-native environments is typically provided through service mesh observability capabilities.

Diagram 9.3 – UEBA Detection Flow in a Zero Trust Environment

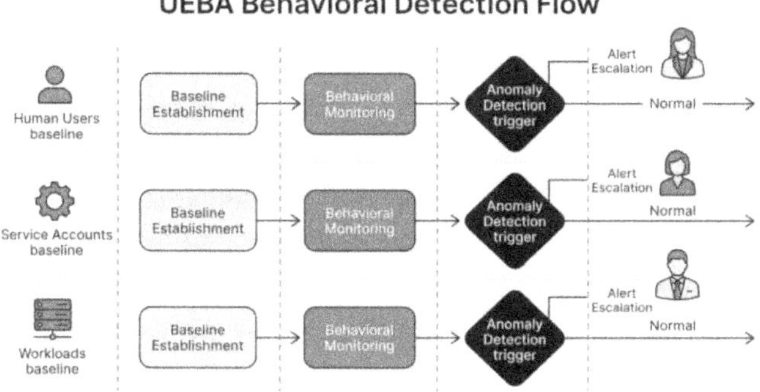

9.2.3 Closing the Loop: From Detection Finding to Policy Update

A detection finding that does not result in a policy change is a missed opportunity to strengthen the Zero Trust architecture. Every confirmed anomaly — whether it reveals an active attack, a misconfigured service account, a policy gap, or a legitimate pattern that the baseline did not anticipate — contains information that should feed back into the policy and monitoring design. The feedback loop from detection to policy is what transforms a Zero Trust program from a deployment into a continuously adaptive security discipline.

The mechanics of this loop require both technical and organizational components. On the technical side, the investigation workflow must include a step that evaluates whether the detected behavior reveals a gap in the policy engine's enforcement logic — an access path that should be controlled but is not, a permission that is broader than the minimum required for the relevant workflow, or a baseline definition that is miscalibrated for the operational context. On the organizational side, the policy operations function must have a mechanism to receive these findings, evaluate them, and update the relevant policies within a defined review cycle.

The speed of this loop matters. A policy gap revealed by a detection finding that takes ninety days to work through the policy review process remains open and exploitable for ninety days after the organization has confirmed it exists. Detection-to-policy timelines should be defined in the governance documentation, differentiated by severity — a gap that enabled an active attack requires immediate policy update, while a gap that revealed a misconfigured baseline can be addressed through the standard review cycle — and tracked as a program metric.

9.3 Governance Structures That Keep the Program Responsive Without Creating Bottlenecks

Zero Trust governance is the organizational mechanism through which all ongoing decisions the program requires — policy updates, exception approvals, maturity target revisions, technology investments — are made consistently, at an appropriate speed, and with the right level of authority. Without governance structures, these decisions are made ad hoc, inconsistently, and at whatever level of authority is available when they cannot be deferred any longer. The result is a policy environment that is technically complex but operationally inconsistent — a patchwork of decisions made under different criteria by different people over time.

The governance challenge specific to Zero Trust is that the program requires decision velocity — the ability to update policies in response to operational feedback, emerging threats, and organizational changes — without sacrificing decision quality. Governance structures that are thorough but slow will become bottlenecks, forcing operational teams to work around the policy process rather than through it. Governance structures that are fast but superficial will produce policy decisions that create new vulnerabilities while closing old ones.

9.3.1 Policy Review Cadences and Exception Management Workflows

Policy review cadences should be differentiated by the type of change being considered. Emergency policy changes

— responses to active threats, critical vulnerability disclosures, or confirmed enforcement failures — require an expedited process that can produce a decision and a deployed policy update within hours. They should not be routed through a standard change management process that operates on weekly or biweekly cycles. The emergency process requires pre-defined authority, a clear escalation path, and a retrospective review within a defined period after the emergency change is made.

Routine policy updates — adjustments to access scopes, the addition of new application contexts, and recalibrations of behavioral baselines — should operate on a defined review cycle, typically weekly or biweekly, with a backlog management process that prioritizes items by risk impact. The backlog should be visible to the full policy operations team and to the security engineering team, so that dependencies between policy items and infrastructure changes can be identified and managed before they create conflicts.

Exception management requires its own workflow, distinct from routine policy updates, because exceptions involve explicit acceptance of risk rather than policy refinement. Every exception to a Zero Trust policy — a device granted access despite failing posture assessment, a user granted elevated privilege outside the standard JIT process, a network path that bypasses a segment boundary — represents a deliberate decision to accept a specific risk for a defined period. That decision should require explicit authority, a defined expiration date, and a review trigger that ensures the exception is re-evaluated before it becomes a permanent feature of the environment.

Exception volume is one of the most informative metrics of governance health available. A low volume of exceptions, well-documented and consistently expiring on schedule, indicates a policy environment well-calibrated to operational reality. A high volume of exceptions, many of which extend beyond their original expiration, indicates either a policy that generates more friction than the operation can absorb or an exception process used as an alternative access governance pathway. Either situation requires management attention, and neither is visible without the tracking infrastructure that exception management governance provides.

Diagram 9.4 – Zero Trust Governance Decision Flow

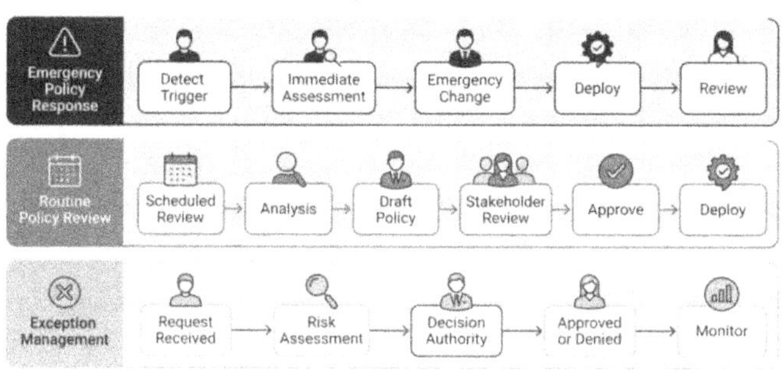

9.3.2 Sustaining Zero Trust Across Leadership and Organizational Change

Every Zero Trust program will eventually face a leadership transition. The CISO who championed the program will leave. The CIO who protected its budget will retire. The executive sponsor who modeled compliant behavior will take a new role. These transitions are predictable, and their impact on the program can be mitigated through governance design that does not depend on any individual's presence.

The key to leadership-resilient governance is institutional documentation. The program's rationale — the risk analysis that justified the investment, the threat scenarios it is designed to address, the metrics that demonstrate its effectiveness — must be maintained in a form that is accessible to incoming leaders who did not participate in the program's design. The governance bodies that make program decisions must have documented charters that describe their composition, authority, and decision processes. The maturity target and the risk acceptance decisions that shaped the program's scope must be documented with enough context to be evaluated and updated by leaders who are seeing them for the first time.

Organizational changes that affect the technology estate — acquisitions, divestitures, major application migrations, cloud platform transitions — each create a specific type of governance challenge: the need to extend or adapt Zero Trust controls to a new organizational or technical context that was not part of the original program design. The governance structure should include a defined process for these integration events: a rapid assessment of the new environment's Zero Trust maturity, a plan to extend or adapt controls, and a timeline that reflects the integration's risk profile rather than the convenience of the schedule.

9.4 Anticipating What Comes Next: Emerging Threats, New Technologies, and Horizon Planning

A Zero Trust program fully focused on maintaining the current architecture against current threats will not be prepared for the changes already visible on the horizon. Artificial intelligence is transforming both the attack surface and the defense toolkit. Post-quantum cryptography is moving from a theoretical concern to an implementation requirement. The regulatory environment is evolving faster than most enterprise security programs can keep pace with. Horizon planning — the deliberate practice of monitoring these developments and assessing their implications for the program before they require an emergency response — is a governance function that most programs underinvest in.

The manager's role in horizon planning is to ensure that the program has a dedicated allocation of attention for this work — not simply to add it to the agenda of the same teams that are occupied with current operations, but to structure time, resources, and authority for the forward-looking assessments that inform multi-year program strategy. This does not require a large investment; it requires a consistent, disciplined practice of monitoring, assessing, and planning that is embedded in the program's governance cadence.

9.4.1 AI-Driven Automation: Opportunities and New Attack Surfaces

Artificial intelligence is creating both significant opportunities and significant new risks for Zero Trust programs. On the opportunity side, AI-driven systems can analyze behavioral telemetry at a scale and speed that no human analyst team can match, identifying subtle anomalies across millions of daily access events that would be invisible to manual review. They can automate the routine policy calibration work that currently consumes substantial capacity from the policy operations team, freeing human analysts for the complex judgment calls that require contextual understanding. And they can accelerate the detection-to-policy loop by suggesting policy updates in response to detected anomalies before a human analyst has completed the investigation.

On the risk side, AI systems introduce new identity types — autonomous agents that make access decisions, execute transactions, and communicate with external services on behalf of human users — that the current Zero Trust architecture was not designed to govern. These agents can accumulate privileges through the actions they are authorized to take, create new data flows that bypass existing monitoring coverage, and be compromised or manipulated in ways that are difficult to detect through conventional behavioral analytics. Governing AI agent identities in a Zero Trust architecture requires extending the identity and access engineering function to address agent provisioning, permission scoping, activity monitoring, and deprovisioning with the same rigor applied to human identities.

The manager's decision in this domain is not whether to allow AI systems in the environment — that decision is being made by business units regardless of the security team's preferences — but how to govern them. An organization that establishes governance for AI agent identities before AI adoption accelerates will be able to extend that governance as adoption grows. One that defers governance until AI agent proliferation has already occurred will face the same challenge it faced with shadow IT: a population of operational capabilities that exist outside the security architecture and that cannot be easily brought inside it without disrupting the business processes that depend on them.

Diagram 9.5 – AI Agent Identity Governance Framework

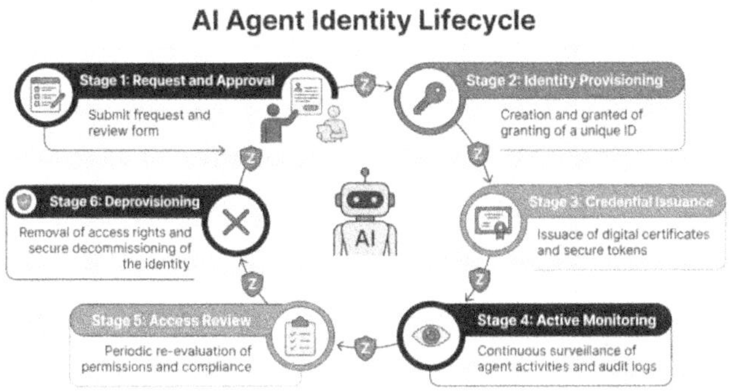

AI Agent Identity Lifecycle

9.4.2 Post-Quantum Cryptography and Its Implications for Zero Trust Controls

Post-quantum cryptography — the development and deployment of encryption algorithms that are resistant to attacks by quantum computers — is moving from a research priority to an implementation requirement on a timeline that most enterprise security programs have not yet incorporated into their planning. The threat model is specific: a sufficiently powerful quantum computer could render current public-key cryptographic algorithms — including RSA and elliptic-curve cryptography — mathematically vulnerable, potentially allowing an adversary to decrypt communications encrypted with these algorithms.

The immediate implication for Zero Trust programs is the "harvest now, decrypt later" attack pattern, in which adversaries are already collecting encrypted traffic from high-value targets with the intention of decrypting it once quantum computing capability matures. Data that must remain confidential for ten or more years — classified government information, trade secrets, patient records — is therefore already at risk from this attack pattern even if quantum computing remains years away from the capability required to execute a real-time attack.

The practical governance question is not whether to deploy post-quantum cryptographic algorithms — that is a matter of when, not if — but how to plan the transition in a way that is sequenced around the organization's highest-risk data first, that accounts for the performance implications of the new algorithms in high-throughput environments, and that coordinates with the technology vendors whose platforms must also support the new standards. The Zero Trust program's encryption infrastructure — TLS sessions, certificate management, and token signing — will all require migration, and the migration plan should be part of the program's five-year governance horizon.

9.4.3 Building a Program That Learns Rather Than One That Endures

The distinction between a Zero Trust program that endures and one that learns is the difference between an architecture that is maintained and one that improves. Maintenance preserves what was built. Learning incorporates new evidence — from detection findings, threat intelligence, operational feedback, and horizon analysis — into a continuously improving architecture that becomes more effective over time rather than simply more experienced.

Building a program that learns requires four structural commitments. First, an explicit feedback loop from monitoring findings to policy design, with defined ownership, timelines, and metrics. Second, a regular horizon-planning practice that translates emerging developments into program-planning decisions before they become operational emergencies. Third, a culture of honest retrospective review — after incidents, after governance decisions, after major deployments — that surfaces what the program got wrong and incorporates those lessons into the next cycle. Fourth, a metrics program that tracks not just the current state of the security posture but its trajectory over time, so that the program can distinguish between improvement and stagnation.

The manager who builds these structures is creating a program that does not depend on any individual's expertise to maintain its quality. The knowledge of what works, what does not, and what is coming next is embedded in the program's processes and governance cadences rather than in the memory of the team members who have been there since the beginning. That institutional resilience is what allows a Zero Trust program to survive leadership transitions, budget constraints, organizational disruptions, and the continuous evolution of the threat landscape that no architecture can outrun through static design alone.

Diagram 9.6 – Zero Trust Program Learning System

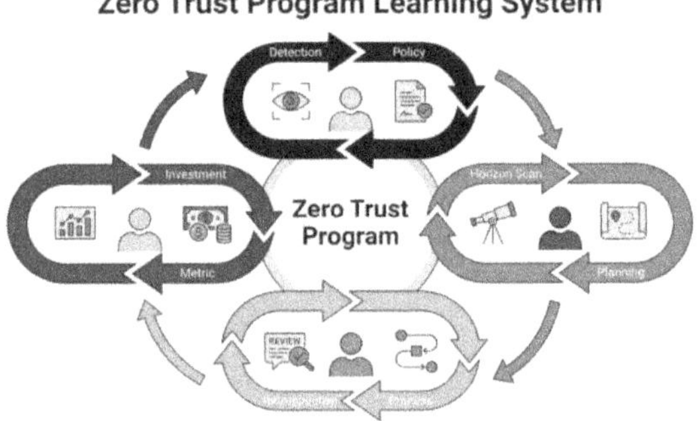

9.5 Manager's Checklist: The Continuous Program

Maintain a telemetry coverage map alongside the enforcement point map — coverage gaps and enforcement gaps are equally exploitable.

Establish a log source signal value assessment that differentiates collection fidelity by the source's demonstrated contribution to detection findings.

Define and enforce tiered retention policies calibrated to detect timelines, investigation requirements, and regulatory obligations — not storage cost alone.

Build correlation capability spanning identity, device, and network signal sources using a common data model to enable cross-domain analysis.

Establish baseline maintenance as a continuous engineering responsibility, with defined processes for recalibrating baselines in response to legitimate changes in access patterns.

Define and track alert fatigue metrics — alert volume per analyst, suppression rate, and time-to-investigate — as leading indicators of the behavioral analytics system's health.

Close the detection-to-policy loop with defined ownership, differentiated timelines by severity, and explicit tracking as a program metric.

Structure governance with three tracks: emergency policy response, routine policy review, and exception management — each with defined authority, timelines, and retrospective review.

Track exception volume and expiration compliance as governance health indicators — persistent or growing exception populations indicate policy calibration or process failures.

Document the program's rationale, governance structures, and risk-acceptance decisions in a format accessible to incoming leaders who did not participate in the program's design.

Establish governance for AI agent identities before AI adoption accelerates—provisioning, permission scoping, monitoring, and deprovisioning with the same rigor as for human identities.

Incorporate post-quantum cryptography migration into the program's five-year governance horizon, sequenced around the highest-risk data assets first.

Build explicit retrospective practices into the program cadence — after incidents, governance decisions, and major deployments — and track what lessons were incorporated in subsequent cycles.

9.6 The Architecture That Never Stops Learning

A Zero Trust program that reaches its initial target state and declares itself completely has misunderstood what it set out to build. Zero Trust is not a project with a completion date. It is an architectural commitment to a discipline of continuous verification that must be sustained, adapted, and improved throughout the enterprise's full lifecycle. The monitoring capabilities, behavioral analytics, governance structures, and horizon planning practices described in this chapter are not supplementary to the architecture — they are the mechanism by which the architecture remains architecture rather than becoming documentation.

The organizations that sustain effective Zero Trust programs over multi-year horizons are not the ones with the most sophisticated initial deployments. They are the ones who have embedded learning into the program's governance structures, who treat every detection finding and every policy exception as information rather than noise, and who have built the institutional resilience to sustain the discipline across the leadership transitions and organizational disruptions that every enterprise will eventually face.

The perimeter model failed because it was static in a dynamic threat environment. Zero Trust is designed to be dynamic — to verify continuously, to adapt policy in response to what it learns, and to extend coverage as the environment evolves — recognizing that design requires not just the technical architecture described in the preceding chapters but also the governance discipline described in this one. Together, they constitute the full program: the architecture and the operating model that keeps it honest over time.

The closing chapter of this book offers a final perspective on what it means to make the architectural commitment that Zero Trust requires — and what that commitment asks of the senior leaders who must sustain it.

10 The Architecture You Choose to Trust

You began this book by examining the collapse of a security model that your organization almost certainly built its infrastructure around. The perimeter that once defined the boundary between trusted and untrusted traffic has dissolved into a landscape of cloud workloads, remote identities, and third-party integrations that no single gateway can meaningfully protect. The nine chapters you have just completed traced a path from that recognition through the principles, components, and operational decisions required to replace implicit trust with verified, policy-driven access at every layer of the enterprise.

The journey moved through identity governance, microsegmentation, and data protection, not as isolated technical upgrades but as interconnected design decisions that reinforce one another. You saw how the control plane and policy engine create the enforcement mechanism, how organizational change and team redesign determine whether the architecture survives contact with real operational pressures, and how continuous monitoring turns a deployment into a living program rather than a completed project. Each of these elements requires sustained attention from leadership, not just funding approval.

The call to action is not to purchase a platform or launch a pilot. It is to make an architectural commitment. Zero trust succeeds when the senior leadership team treats it as a governing principle for how the enterprise manages access, validates identity, and protects data across every system and every interaction. That commitment must be visible in budget allocation, staffing decisions, performance metrics, and executive communications. Without that visibility, the program will stall at the boundaries of the first team that decides the verification requirements are inconvenient.

Start with the credential inventory you have been deferring. Identify the three systems in which implicit trust creates the greatest exposure. Assign an owner to the first phase of microsegmentation. Set a review cadence that survives the next leadership transition. These are not aspirational suggestions. They are the minimum viable actions that separate an organization that understands zero trust from one that is merely aware of it.

The architecture your enterprise trusts today was designed for a threat landscape that no longer exists. The architecture you build next will define whether your organization controls its security posture or merely reacts to the next breach. That decision belongs to you, and the framework for making it is now in your hands.

11 Conclusion

When the Machines Began to Move Faster

There is a moment, in every era of technological change, when the world shifts quietly beneath our feet. Most people don't notice it at first. The dashboards still light up. The meetings still happen. The systems still hum. But somewhere in the background, something fundamental has changed. The rules that governed yesterday no longer apply, and the assumptions that once felt safe begin to crack.

For cybersecurity leaders, that moment has already arrived.

Across this book, we have walked through a landscape transformed—sometimes violently—by artificial intelligence. We have seen attackers who no longer wait, who no longer guess, who no longer rely on human patience or human error. We have seen malware that learns, phishing campaigns that speak in familiar voices, and synthetic media that can fracture trust with a single fabricated sentence. We have seen defenders racing to keep pace, building systems that analyze behavior, triage alerts, and respond at machine speed. And we have seen organizations grappling with the uncomfortable truth that the threat is no longer "out there." It is woven into the very technologies that power modern enterprise.

But beneath the technical detail, beneath the diagrams and frameworks, something deeper has been unfolding: **a**

story about leadership in a world where intelligence—human and artificial—collide.

This is not a story about tools. It is a story about people.

It is the story of a SOC analyst staring at a screen full of alerts, knowing that somewhere in that noise is the one signal that matters. It is the story of a CISO walking into a boardroom, carrying the weight of a threat landscape that changes faster than budgets do. It is the story of a federal program manager trying to secure a system built twenty years ago against an adversary built twenty minutes ago. And it is the story of every leader who has ever felt the quiet pressure of responsibility—the knowledge that the decisions they make today will determine whether their organization survives tomorrow.

AI has not made that responsibility easier. It has made it heavier.

But it has also made it more meaningful.

Because in a world where attackers move at machine speed, the organizations that endure will not be the ones with the biggest budgets or the most tools. They will be the ones with leaders who understand what this moment demands: clarity, courage, and a willingness to rethink everything they thought they knew about security.

- Leaders who understand that governance is not bureaucracy—it is the architecture of trust.
- Leaders who understand that talent is not a cost—it is the engine of resilience.
- Leaders who understand that adaptation is not a reaction—it is a discipline.

- Leaders who understand that AI is not a threat or a savior—it is a force multiplier, shaped by the hands that wield it.

The future of cybersecurity will not be written by machines. It will be written by the people who learn how to lead alongside them.

As you close this book, imagine the enterprise you are responsible for. Imagine its systems, its people, its mission. Imagine the adversaries who are already probing its edges. And then imagine the version of that enterprise that is ready—not just for the threats of today, but for the ones that have not yet been invented.

That future is not built in a single decision. It is built in a thousand small ones:

- A budget reallocated toward automation.
- A governance framework updated before it becomes necessary.
- A training program launched before the skill gap becomes a crisis.
- A culture shaped intentionally, not accidentally.
- A team empowered to think, question, and adapt.

This is the work of leadership in the age of intelligent threats. It is not glamorous. It is not simple. It is not quick. But it is necessary. And it is possible.

The machines may move faster now. But leadership still determines the direction.

And in that truth lies the defender's greatest advantage.

The next chapter of cybersecurity will be written by those who choose to lead—not from fear, not from compliance, but from conviction. The conviction that resilience is achievable. That trust can be rebuilt. That teams can evolve. That organizations can adapt. And that intelligence—human and artificial—can be harnessed not just to defend the enterprise, but to strengthen it.

The age of intelligent threats has arrived.
So has the age of intelligent defense.
The question that remains is simple:
What kind of leader will you choose to be?